Quiet Activism

"Alongside the depressing reports of climate breakdown, *Quiet Activism* is a book you must read. It is an essential reminder that we need to approach each other with compassion, use local knowledge and build an ethics of care if we are to build meaningful and transformative climate action in these troubling times".
—Paul Chatterton, *Professor of Urban Futures, School of Geography, University of Leeds, UK*

"At last, a book that carefully explores everyday practices and the local as spaces of transformative potential in climate activism. The focus on quiet forms of change, creative disruption, generative acts, small scale collectivity and attention to care is inspiring and yet also pragmatic. This book is full of hopeful possibilities and examples of people responding to the climate emergency".
—Jenny Pickerill, *Professor of Environmental Geography, Department of Geography, University of Sheffield, UK*

"This is an important book with a vital and persuasive argument. It makes the case for the power of collective local action, providing compelling examples of a climate activism that is modest, caring and socially innovative, building enduring commitments and pathways for transformation that transcend the individual. Quiet activism is not enough on its own, but this book makes absolutely clear why it should not be dismissed or undervalued".
—Gordon Walker, *Lancaster University, Lancaster, UK*

"The authors do a great job of highlighting the critical need for *Quiet Activism*: the intimate, embodied acts of collective disruption, subversion, creativity and care that individuals and groups are practicing at the local scale. In doing so, they convincingly foreground the transformative power of socially innovative activities and initiatives in response to the brutal realities of our climate crisis."
—Professor Julian Agyeman, *Tufts University, Medford, Massachusetts, USA*

Wendy Steele
Jean Hillier • Diana MacCallum
Jason Byrne • Donna Houston

Quiet Activism

Climate Action at the Local Scale

Wendy Steele
RMIT University
Melbourne, VIC, Australia

Jean Hillier
RMIT University
Melbourne, VIC, Australia

Diana MacCallum
Curtin University
Perth, WA, Australia

Jason Byrne
University of Tasmania
Hobart, TAS, Australia

Donna Houston
Macquarie University
Sydney, NSW, Australia

ISBN 978-3-030-78726-4   ISBN 978-3-030-78727-1   (eBook)
https://doi.org/10.1007/978-3-030-78727-1

This Palgrave Macmillan imprint is published by the registered company Springer Nature Switzerland AG.
The registered company address is: Gewerbestrasse 11, 6330 Cham, Switzerland

*To all those working in their communities at the local scale
to involve, engage and inspire people to act "now" to address
the climate emergency
— you are not alone.*

ACKNOWLEDGEMENTS

We would like to acknowledge that we live and work on the unceded lands of Aboriginal and Torres Strait Islander people, and as uninvited guests, we pay our sincere respects to the Traditional Custodians and Elders past, present and emerging.

This project builds on work that the project team have been developing over a number of years—collectively and separately—in the fields of climate justice, environmental action, geography, urban planning, politics and policy and place-based social innovation analysis. Central to this agenda is a shared interest in how we can re-imagine our experience of, and responses to, the climate-changed city. We see action on climate change as an urgent, cross-disciplinary and cross-sectoral whole-of-society agenda. Our emphasis is on climate action as the "wilful hope" and "care-full practices" of everyday activism at the local scale. This for us is quiet activism—the socially innovative, modest yet powerful acts of localized activity, care and community. We argue in this book that quiet activist practices offer up powerful pathways for local communities to act hopefully in the face of the climate emergency.

The award of an Australian Research Council Discovery Grant (ARC DP150100299) on "Socially innovative adaptation to climate change at the local scale" provided an opportunity to extend and deepen our research within the Australian context. The project was led by RMIT's Centre for Urban Research (CUR) in partnership with Macquarie University, Curtin University, The University of Tasmania and Griffith University. The research investigated the practices and framings of community groups, NGOs and local governments in the creation of local adaptation strategies. Through case-studies, interviews and focus groups in Melbourne, Sydney, Brisbane and Perth culminating in a national stakeholder forum and public facing website (see climateadaptationaustralia. com.au), the ambition of the project was to use these understandings to direct public and policy attention towards the need for building social innovation and capacity at the local scale in response to the climate change crisis.

At the heart of the research were the many participants who generously contributed their time and experiences. As co-producers of knowledge and learning, we are deeply grateful for the passion, creativity and commitment of the participants in our online survey, interviews and focus groups from across community, government and the private sectors. This includes: all of the presenters at our national stakeholder symposium including Fiona Armstrong, Guy Abrahams, Jo Bower, Ken Baird, Robert Enker, James Duggie, Geoff Love, Dorean Erhardt, Bill Forrest, Dianna McDonald, Shani Graham, Phil Ingamells, Ryan Quinn, Greg Hunt, Joseph Kelly, Pradesh Tamiah, Griff Morris, Karl Mallon, Paul Sirant and Rory Martin; other individuals highlighted in the book including Ann Noble, Alan Pears, Cate Ware and Tim Darby and organisations across sectors and Australian states including Marrickville Council and Renew (the Alternative Technology Association), the Elwood Floods Action Group (EFLAG), RedWaste, Climate for Change, Green Cross Australia, Environment House and the *ReNew* Initiatives, GECKO, Climarte, One Planet, CANWin, the Mossvale Community Garden, Darebin Solar $aver and Ecoburbia. To everyone who participated in this research project—thank you!

Research is an extended community of entangled practices, and in this we were fortunate to have Katelyn Samson as our project manager extraordinaire, and research associates and assistants Mary O'Halloran, Tom Overton Skinner, Ryan Quinn and Jayden Holmes. We also benefited greatly from the support of our expert reference group of Professor Robyn Dowling, Doug Perdie, Dr. Michelle Maloney, Huxley Lawler, Brendan Sydes, Campbell Watts, Melanie Bainbridge and Irina Cattalini chaired by Professor Hartmut Fuenfgeld; and the concomitant research of Dr. Lisa de Kleyn who completed her PhD offering a situated approach to environmental justice in Toolangi State Forest funded by the project.

CONTENTS

About the Authors

Wendy Steele is an award-winning researcher, writer, activist and educator in the Centre for Urban Research at RMIT University, Melbourne, Australia. Her recent books include *The Sustainable Development Goals and Higher Education: A Transformative Agenda?* (Palgrave 2021), and *Planning Wild Cities: Human-nature relationships in the urban age'* (Routledge, 2020).

Jean Hillier is an emeritus professor at RMIT University whose research interests include post-structural planning theory and methodology for strategic practice in conditions of uncertainty, political and cultural aspects of governance activity and more-than-human planning theory and practice.

Diana MacCallum is Adjunct Academic in Urban and Regional Planning at Curtin University. Her research focuses broadly on social aspects of planning and development. She has co-authored or edited six books, including *The International Handbook on Social Innovation* and *Advanced Introduction to Social Innovation*.

 **Jason Byrne** is a Professor of Human Geography and Planning at the University of Tasmania. He researches urban political ecologies of green-space, climate change adaptation and environmental justice. Jason has previously been awarded the Planning Institute Australia's national award for cutting-edge research and teaching.

 **Donna Houston** is an urban and cultural geographer in the Department of Geography and Planning at Macquarie University. Her research explores the intersections of urban political ecology and environmental justice in the Anthropocene, cultural dimensions of climate change, spaces of extinction and planning in the "more-than-human" city.

LIST OF FIGURES

CHAPTER 1

Addressing the Climate Emergency at the Local Scale

Abstract *This chapter* introduces "quiet activism" and its relationship to social innovation, adaptation and other forms of activism with an embodied ethic of care. This includes an emphasis on the importance of the local scale in addressing the climate emergency and other crisis; cases of socially innovative practices and partnerships from Australia alongside international examples; and critical engagement with the role of quiet activism as a transformative conceptual frame and contemporary practice in climate change. We are interested in the diverse ways this transformative potential of quiet activism can create more equitable and sustainable futures.

Keywords Quiet activism • Social innovation • Climate emergency • Sustainability • Local

WHY QUIET?

As I write humanity stands at a fork in the road. Unless we act decisively to phase out the use of fossil fuels, global temperatures will exceed a 2 degree rise above the pre-industrial levels in a few decades, and we will risk committing every human to climatic shocks and catastrophes that will destroy our civilization and precipitate mass extinctions.[1]

This book focuses on the potential and possibilities of socially innovative community responses to the climate emergency at the local scale. We

© The Author(s), under exclusive license to Springer Nature Switzerland AG 2021
W. Steele et al., *Quiet Activism*,
https://doi.org/10.1007/978-3-030-78727-1_1

are particularly interested in co-produced forms of local climate action which we describe as "quiet activism". The creative and collaborative ways local-scale climate action reflects the 'extra-ordinary measures taken by ordinary people'.[2] Climate change has intensified the need for communities to find meaningful and transformative ways to better address the sustainability of their environments. This includes critical engagement with how and in what ways novel social practices and partnerships emerge between people, organisations, institutions, governance arrangements and ecosystems. The aim of this book is twofold: to highlight the transformative power of socially innovative activities and initiatives in response to the climate crisis; and to critically explore how different individuals and groups undertake climate action as quiet activism—the intimate and embodied acts of collective disruption, subversion, creativity and care at the local scale.

There has been a longstanding feminist critique of normative visions of activist behaviour as vocal and antagonistic forms of protest versus more modest, embodied acts of care, connection and creativity as part of everyday life, places and spaces. A range of literature in a range of contexts (although typically feminist) has sought to draw attention to the power of small, purposeful everyday practices of resistance and rebellion, the politics of making and doing, and the ways in which this produces both the means and conditions through which alternative values can be explored and shared.[3] Much of the focus in this diverse body of work is on the local, informal, socially engaged nature and deployment of subversive strategies and tactics: for example women's work at home, and work in the academy to support and advocate for women's rights and reconfigure the nature of "work";[4] or creative activity and engagement with local placemaking, equity and sustainability issues through guerrilla gardening, yarn bombing, counterfeit crochet, punk DIY, pop-up initiatives and tactical urbanism.[5]

From a very different perspective comes Deborah Bird Rose's (2004) powerful description of decolonisation in *Reports from a Wild Country* set within the settler-colonial history (and present) of Australia. She describes the need for practical, localised engagement with the "here and now" as an ethical encounter and politics of entanglement with past, present and future. Rose highlights the description of the invader Captain Cook by Hobbles Danayarri, a now deceased Yarralin lawman and community leader. 'As Hobbles liked to say, Captain Cook was the real wild one. He failed to recognize the Law, destroyed people and country, lived by damage, and promoted cruelty'.[6] Rose offers up a situated understanding based on her discussion with Indigenous elders. '*Quiet* country stands in contrast to the wild: man-made and cattle-made'.[7]

Climate change as a pathway towards reconciliation and regeneration also invites us to bear witness "here and now" to the unsustainable nature of current development trajectories and the on-going legacies that co-constitute modernity's "developmentalities".[8] This depends on moving beyond the illusions of a comfortable life to being present (a witness) to the moral claims being offered without retreating to a position in which the 'current contradictions and suffering will all be left behind justified by references to the future'.[9] A future orientation, Rose argues within the context of settler-colonialism and climate change, 'has been a major tool deflecting us from moral responsibility'.[10] An alternative approach is a regenerative politics that seeks to re-shape the nature of relationships between people, place and environments from the ground up. The emphasis is focused instead on the localised knowledge and capabilities that lead to transformative change, without defaulting to a distant future imaginary, or an escape from everyday accountabilities and responsibilities.

This book offers conceptual and practical insights into the challenges and opportunities of quiet activism as a local-scale response to the climate emergency. In particular we critically address the tendency to frame "quiet" as being at best a precursor to political action, and at worst disempowering or conservative—reinforcing a neoliberal agenda of individualised action and responsibility. Our focus is on the transformative potential of local, socially engaged, subversive strategies and tactics, and the ways in which these can help to create new social relations and different urban imaginaries. The intention is not to privilege or polarise quiet activism above or against other forms of social movement, political protest or climate activism. Instead, our aim is to highlight the need for, and importance of, a diversity of climate *activism/s*, and reinforce the vital role of "quiet activism" within this: the co-produced nature of small-scale disruption, ingenuity, creativity and political craft-building already underway at the local scale to address the climate emergency.

WHY LOCAL?

In the trifold crisis of climate change, Coronavirus and the Black Lives Matter (BLM) movement the need for transformative societal change has refocused attention on the importance of the local scale. This is where many people see and feel (i.e. "live") the multiple effects of climate change and crisis. These everyday experiences and encounters unfold in local places that are important sites for thinking about how the bigger stories of change and disruption are experienced, lived, negotiated and responded to. The innovativeness and local appropriateness of local scale approaches to action on climate change tend to be undervalued or discounted.[11] Given we now live in a climate emergency, surprisingly little is known about why this occurs and what alternatives exist.

The task of addressing climate change is most acutely felt at the local scale to reduce vulnerability and build social innovation and community capacity in response. This is where international, national and state policies are translated into on-the-ground practices that help people and places better prepare for, adapt to and mitigate against the impacts of climate change (e.g. sea level rise, flooding, heatwaves, higher intensity storms, etc.). Government authorities, tasked with leading local adaptation, are often caught in an implementation trap. Responsible for issues of land use and environmental management, local governments can face daunting investor/constituent pressure and potential litigation over decisions associated with land zoning and development. Governments can become caught in an impasse where they are simultaneously both taking action on climate change and being risk-averse, seeking to avoid decisions and actions that may heighten their exposure to future liability.

Effective action on climate change is challenging governments where formal responses to climate change are dominated by technical risk management approaches.[12] There are many examples of local governments declaring "climate emergencies" and doing lots of diverse work at a very local level, not just technically but involving communities in gardens, solar power schemes and so on. These approaches can have great merit but adopting generic risk-assessment tools can lead to the development of plans that fall short of meeting local concerns and fail to effectively incorporate local knowledge and skills.[13] For example, risk-based plans are resource intensive and fixed in technical analysis. Such plans can be expensive to implement and require specialist expertise making them difficult to apply to changing circumstances, especially if municipal budgets are strained.[14] Whilst many local governments have undertaken risk assessments and know the challenges they face, they are often unable and/or unwilling to take the action needed to reduce carbon emissions quickly and implement climate mitigation and adaptation initiatives that are innovative and equitable.[15]

In this book we focus on local-scale stories that demonstrate how barriers to action on climate change have been broken down, and where new opportunities for collective action have been identified and developed. Effective responses to climate change involve local practices in governance, civics and science-community-industry collaborations that work to challenge the current post-political environment and highlight the different pathways and portals needed to address the climate emergency. We seek to draw critical attention to both the localised responses to the climate change emergency that seek to support and promote environmental sustainability, and the ways in which we can learn from different modes of community action and resistance to create new knowledge that supports and sustains social innovation and capacity building at the local scale.

Local practices are situated in particular contexts and circumstances, and grounded in particular places and spaces through emerging partnerships, initiatives and actions. They do not adhere to any one model or outcome. In this regard, these diverse and creative local climate actions, initiatives and experiments are contingent and open to continual, collaborative work. A common thread in this book is a focus on what people care about, and how "care-full" community practices are located within everyday practices of connected reciprocity and responsibility to address climate change. The diverse ways of interweaving climate action at the local scale raise questions such as what it means to be an activist, or "do activism"; and the implications for building greater capacity for sustainable living are foregrounded. This contrasts with other types of action on climate change such as the *Extinction Rebellion,* which seeks to deploy a very different kind of activism as part of a mass global movement.

The Role of Social Innovation

Given the limitations of existing technology *and* institutions to respond adequately to climate change (and other grand challenges of the twenty-first century), "innovation" has become a key buzzword in policy and academic circles.[16] Contradictory pressures can lead to inertia and create gaps in meeting human needs; however, these gaps—while often exclusionary—can also constitute spaces in which innovation can thrive;[17] "lines of flight", in Deleuzian terms,[18] for emergent transformational potential. More specifically, since the early 2000s we have observed a rapidly increasing interest in *social* innovation.[19] While we recognise that this is a contested concept,[20] we use it here to acknowledge that 'innovation in social relations' is just as important—if not more important to confronting the threat of climate change,—as technologies and new investment opportunities.[21]

Social innovation is an alternative approach to the constraints that risk management appears to place on the scope for adaptation responses and their implementation. This type of innovative practice typically entails the creation of new organisational structures, service delivery modes, products and activities that efficiently and effectively meet social needs or provide social benefits by grounding them in the relations and experiences of excluded groups.[22] In our research we see the potential of social innovation as a form of empowerment that fosters and enables new forms of cooperation to provide for previously unmet needs. The practices associated with social innovation are not only "innovative" in creating solutions to otherwise intractable problems, but also more inclusive and empowering than traditional approaches, enabling new actors to contribute to solutions.

However, the role of socially innovative practices—particularly within the context of action on climate change at the local scale—is poorly understood. Social innovation in practice does not separate institutional means from material ends. Rather, it responds to needs (whether material, cultural, ecological, etc.) unmet by state and market through the adoption of new/experimental social practices and institutional forms that provide for democratic and solidarity-based processes.[23] Socially innovative strategies thus can lead to *collective* action; and can create the conditions for broader and/or deeper social change.[24]

In our research on climate action at the local scale within the Australian context, the participants all recognised the centrality of social relations to a liveable future in a changed climate. They are aware of the importance of inter-scalar dynamics between the local, where those relations are shaped by proximity, and the effects of crisis are most clearly manifest; and the state and the global, where the political-economic causes of the crisis reside. Their initiatives attempt to navigate these dynamics in socially innovative ways in the face of tensions and contradictions which, though quite specific to each place and organisation, have striking thematic resonances with what we describe in this book as "quiet activism"—the everyday practices undertaken by ordinary people to address the climate emergency. Our research focus was the suburban heartlands of Australia's major cities.

The Australian Research Context

Australia is one of the world's driest continents and with a population perched precariously in coastal cities highly vulnerable to sea level rise it is a global urban hotspot at the frontline of climate change responses within the developed world.[25] It is also a settler society, like the United States, Canada, South Africa, and thus offers insights into global social phenomena that complicate and often frustrate effective climate change responses at all scales (e.g. post-colonial politics, globalisation and race-relations). In *Surviving* Julianne Shultz describes that 'the geographic diversity of the island continent means that nature's extremes take different forms, depending on latitude, longitude and patterns of settlement—fires for some, floods and cyclones for others, elsewhere earthquakes and storm surges or the lingering, prodding fingers of drought'.[26]

The highly concentrated nature of the urban population coupled with the relatively fixed nature of much of the metropolitan built form serves to magnify climate-related risks and vulnerabilities from extreme weather and natural disasters (i.e. sea-level rise, heatwaves and drought). The impacts of climate change compound existing urban vulnerabilities such as poverty, inadequate public infrastructure, loss of biodiversity or environmental degradation. Yet the responses to these issues are locked into processes that are largely hardwired into dry economically unsustainable models of growth and development experienced most acutely at the local scale. Within the Australian context, in the absence of national policy and leadership on climate change,[27] it is at the local scale that innovative and engaged action is quietly creating the conditions and sowing the possibilities for much-needed societal change.

In the early twenty-first century Australia, like many other countries, is experiencing extreme heat, severe bushfires, intense flooding and drought conditions at the same time. This mirrors recent global events, which included intense hurricanes and wildfires (the United States), severe drought (Argentina, South Africa), extreme heat (the United Kingdom, Norway, Finland), fires (Greece, Sweden) and extensive storms and/or flooding (Germany, Japan, Indonesia), to name but a few. The compounding impact of multiple events serves to make climate disasters worse: bushfires are intensified by tornadoes, storms and lightning strikes, landslides are caused when rain hits the charred remains of fire-burnt regions.[28]

A report by the Climate Council *Weather Gone Wild* emphasises that extreme weather events such as bushfires are part of a trend of increasingly frequent catastrophic weather events in Australia and globally. The report highlighted four key findings: that recent years have been the hottest on record for global surface temperature continuing a long-term warming trend; climate change is increasing the frequency and/or severity of extreme, wild weather both globally and in Australia; the impacts of extreme weather have been damaging and costly; and to slow and eventually stop the increase in the frequency and severity of extreme weather, Australia needs effective climate policy and action to drive down greenhouse gas pollution deeply and rapidly.[29]

To explore climate action at the local scale, our research funded by a three-year Australian Research Council (ARC) grant employed a critical social science review of the international literature and a practice-based approach to local-scale case examples in Australian metropolitan areas at the forefront of the climate change crisis. The research presented in this book focuses on the nexus of, and relationships between, social innovation (broadly defined as innovative practices operating in both formal and informal contexts) and local climate action. We are particularly interested in everyday, local-scale climate initiatives and activities that seek to meet genuine needs, engage and empower the local community in the preparation and delivery of strategy and initiatives, hold the potential for the transformation of social relations, and prioritise those most vulnerable including non-human species.

Central to our research is an interrogation of the tensions and potentialities between top-down risk assessment by agencies of local governance, bottom-up innovations by local activist groups and non-government organisations, as well as the spaces in-between. The object of research inquiry for this project focuses on: how different actors at the local scale situate themselves in relation to action on climate change; seek to address it; interact with each other; and generate (mal)adaptive responses. We investigated the framings and practices of local governments, community groups and non-government organisations as they sought to create local strategies reflecting lived realities at the local scale (e.g. operating budgets, character of housing stock, aesthetic preferences, political conflicts and physical characteristics of local environments). Our research explores the practices of what we describe as "quiet activism", as both a prefigurative and regenerative politics.

Four key questions guided the initial research including:

1. How do citizens, political decision-makers, policymakers, planning officers and service and advocacy groups frame climate action at the local scale, and do different framings lead to conflict and/or cooperation?
2. What institutional and cultural forces shape local actors' understandings, framings and practices, and how do local actors respond to these forces, especially across scales?
3. What factors determine whether the different framings and practices they engender translate into policy decisions and on-the-ground climate action?
4. How do local actors 'go round the back' of local institutions' mainstream approaches in search of socially innovative responses that better meet their needs?

To address these questions, we involved a review of the international literature as well as five practice-based research phases which included the establishment of a Project Reference Group; a comprehensive audit of local governments and identification of case studies; detailed investigation and analysis of case studies; in-depth interviews and focus groups; and participant observation of selected local-scale innovative initiatives addressing climate change in action. Our focus was on the ordinary acts of care that constitute everyday action on climate change as regenerative practices at the local scale in Australia's capital cities.

To this end we were not explicitly seeking the experiences of those most marginalised and on the fringe in Australia's capital cities: the edge dwellers of the (sub)urban periphery such as refugees and the homeless. Nor did we work closely with First Nations communities on whose unceded land Australian metropolitan areas are founded and where the research was "emplaced". This work is already underway. The First Nations climate-justice organisation *SeedMob* is a flourishing network supporting frontline communities and Indigenous leadership in the youth climate movement.[30] *SeedMob* addresses the intersections of settler-colonialism and climate injustice and centres First Nations' cultural resurgence and caring for country for the benefit of all.[31]

At the time of our project, the combined sampling process of self-selection and voluntary participation resulted in the majority of participants being predominantly white, Western and middle class. Our commitment to climate justice, critical reflexivity and ethics-based research practice infused the project throughout, however we recognise the contradictions and tensions of our unearned privilege as white-settler researchers. Our research shows the persistence of settler values in local climate action in Australia. This reinforces First Nations' author Tony Birch's call for white Australia:

> to address a question. How does the nation move from a state of colonial anxiety that refuses genuine recognition and engagement to a concept of locating 'Indigenous theories, methodologies and methods at the centre, not the periphery of our society? While such a shift could ultimately produce 'an ecological philosophy of mutual benefit', getting there will be a real challenge.

We acknowledge this challenge and the limitations of the research presented here. Our hope is that in attending to quiet climate activism at the local scale, greater capacity for dialogue, engagement and material action on the vitally important work of centring First Nations histories, places and knowledge in the climate emergency are opened up.

A key element of the project was the formation of a stakeholder Reference Group, designed to act as a "critical friend" to provide a conceptual and practical sounding board for the life of the project. Involvement of stakeholders in the research process is an important development which can ensure that the research aligns with lived experience and produces outputs that have practical utility. The Reference Group included academia, local government, state government, non-profit and/or social enterprise and the private sector. We met in person at the beginning, middle and the final phases of the project, and electronically twice per year to provide advice on key decisions such as selection of case studies, integration of actors from different sectors and development of new models for practice.

The project also mapped the current existence and status of climate change strategies in local governments across Australia who were asked to identify what they perceived to be successful and/or unsuccessful climate action in their area and why. This provided data relating to local framings and practices in local planning. Analysis focused on the key indicators, conditions and parameters of "success" and implementation from a local government point of view. This was supported by a simultaneous audit of community-based websites to identify socially innovative organisations and locally based climate action projects. Three types of action were explored and developed in further detail: awareness raising, policy drivers and gap-bridging.

Building on this scoping phase, a detailed investigation and analysis of case studies focused on how practices associated with climate adaptation responses shape lived outcomes. The in-depth case investigations of local government and community-based initiatives were undertaken over several years and attended to several related elements/data sources, each of which expresses the discourses and practices of climate action at the local scale. These include published texts such as plans, strategies and "official" documents; in-depth interviews and focus groups with a range of actors involved with the development and implementation of adaptation plans and projects, including representatives from state and local government, NGOs, social enterprises and the private sector; and ethnographic observations to understand practice regimes, as it is necessary to observe them in operation.

The analysis focused on teasing out issues, concerns, frictions, constraints and possibilities related to these processes; critically exploring the challenges and opportunities for climate action, including where plans have been abandoned or "put on hold"; and responsiveness to local conditions such as social inclusion, institutional capacity, identification of embedded (human and other) resources for adaptation, place making and empowerment of local communities to act. These tensions are critical to understanding how practices frame and shape outcomes within local settings and provide the framework for what we describe as "quiet activism" outlined in this book and summarised in section below.

Towards a Quiet Activist Framework

Our emphasis in this book is bringing together the ideas and practices that shape the activities that support the action on climate change. Building on social innovation and critical practice theory, these include the localised stories of climate action that make the connections between people and the material and natural world; the benefits or detriments of the climate-just city for marginalised communities within particular urban contexts; how addressing the climate emergency challenges, compliments or replaces current rights and existing privileges in cities; and how climate action incorporates ethics and responsibilities that recognise the ways both humans and non-humans participate in climate activism and participatory governance.

We are particularly interested in how climate action is framed by local governments, the private sector and NGOs in ways that are both "innovative" and "ordinary" as mutually reinforcing activist agendas in everyday local urban contexts. Based on this, we identified five key practices that might be put to work in shaping local-scale climate action. The following five key themes emerged from our research and together underpin an emerging framework for understanding "quiet activism": building and bridging the knowledge base; bringing missing actors to the table; walking together with care; making and breaking connections; and realising transformative potential (see Fig. 1.1).

Fig. 1.1 Quiet activism: a framework for practice. (Source: Authors)

1) *Powerful stories: Building and bridging the knowledge base*

At a local scale this range of ways of "knowing" involves a spectrum of stories: informal, subjective accounts such as dialogue, personal experience, local knowledge, visual images, actions and activities; as well as the more formal, objective reports such as resource mapping and scientific research. Stories in all their forms are the compasses by which we navigate our place in the world, offering a way of travelling from here to there and locating the silences in between. Stories help build knowledge through the narratives, ideas, anecdotes, summaries, histories, musings, theories and data that work to create empathy and identity. There is a rich "multiplicity of ways of knowing"[32] or "connected knowing".[33] Building and bridging the knowledge base involve engaging with diverse ways of knowing through a constellation of stories, experience and activities—the innovative quiet activist practices and relationships that take root and flourish at the local scale.

2) *Response-ability: Bringing missing actors to the table*

Central to our analysis is the idea of "enabling" innovative practices[34] as a form of quiet activism. Innovation might be conceptualised as occurring across three domains—behaviour change, institutional change and built form change—such as tree planting, lighter coloured roads and roofs, heatwave shelters, stormwater harvesting and so on. Elements of enablement include leadership, relationship building (e.g. trust, commitment, open communication, collaboration and information exchange), skills development and competence building (e.g. problem identification, developing solutions), decision-making support (e.g. financial or human resourcing, delegating power to act) and response-ability (i.e. to external disruptors, e.g. storm events or financial shocks, risk management, entrepreneurial readiness, experimentation).[35] Quiet activist practices that map onto these dimensions of enablement illustrate how each dimension offers a different way to bring missing actors to the table.

3) *Quiet adaptation: Walking together with care*

The multiplicity of urban weather-worlds forms significant eddies and undercurrents where governments, individuals, groups and organisations trying to implement climate adaptation are pulled in many different directions, leading to more fragmented rather than genuinely collaborative approaches. The focus by governments and organisations on creating a consensus around climate change represents a "loud" form of climate adaptation that drowns out the smaller, more contingent, less certain and more experimental forms of adaptation that can quietly generate meaningful action. The real and perceived requirement to overcome the unpredictable contingency of climate change via forms of public or political consensus can be a destabilising and de-politicising force, which refuses a diversity of experiences and practices, including the capacity to see where climate adaptation practices may be creating their own forms of harm.[36] Quiet activism involves walking together with care, and in doing so being more attuned to the spatially uneven, everyday worlds and practices that people inhabit and are complicit in making and remaking.

4) *Scaling out, up and deep: Realising transformative potential*

Different types of actors come into contact through different types of technologies, using different systems and modes of communication, and responding to opportunities that resonate with their different concerns and daily practices. Important to "scaling out" are various modes of influence, which allow actors not associated with the original actions to adapt and replicate them in other places. The processes of "scaling up" are vitally important, meaning the higher level institutionalisation of innovation by way of new or reinterpreted policy and public governance arrangements—that is, changes to the political context in ways that allow local experiments to sustain themselves and to flourish. Finally, new practices change not just organisations, but people too. This is "scaling deep"—having a profound effect on the lives and minds of the people who participate in an initiative. It means that everyday activist practices can become embedded as a way of life—a new normal.

5) *Critique, subvert, rework: Making and breaking connections*

The transformation of practices requires a multilevel perspective which highlights the connections, or points of intersection, between scale and practice.[37] Three tiers include *The broader socio-technical landscape*—the wider context which includes not only technical and material elements, such as the electricity grid, laws and regulations, international standards of thermal comfort, capitalist economics, but also demographic trends, societal values and political ideologies; *Socio-technical regimes*—comprise shared beliefs, established practices and rules, institutional ways of defining problems and so on, which stabilise existing systems and infrastructure; and *Niches*—not market niches, but rather like evolutionary niches in biology, are innovations, experiments and novel ideas which deviate from existing regimes, such as renewable energy technology in the home.[38] Quiet activism as the practice-based connections of images, skills and materials are also horizontal, such as between local interest groups, between committee members or across regimes.

LOCAL LINES OF FLIGHT

How might we re-imagine our experience of, and responses to, the climate emergency? A critical practice-based approach highlights a diverse range of local-scale, socially innovative activities and initiatives focused on responding to climate change; and the emerging conceptualisation and practices of "quiet activism" and its relationship to social innovation, adaptation and other forms of activism with an embodied ethic of care. This includes an emphasis on the importance of the local scale in addressing the climate emergency and other crises; cases of socially innovative practices and partnerships from Australia alongside international examples; and critical engagement with the role of quiet activism as a transformative conceptual frame and contemporary set of practices in climate change.

This introductory chapter sets the context and structure of this book, including the practice-based approach to research. The emphasis on socially innovative climate action focuses attention on the experimental partnerships and collaborative learning activities that are collectively and politically significant at the local scale. Creatively and purposefully participating in local activities lays the foundation for new socially innovative practices in response to climate change. This is the power of small, purposeful everyday practices of resistance and rebellion, as both the means and conditions by which altered values can be explored and shared through care, connection and creativity at the local scale. We are interested in the diverse ways this transformative potential of quiet activism can create more equitable and sustainable futures.

In the following six chapters we draw specific conceptual and empirical attention to each of the five themes we have identified in the Quiet Activism Framework highlighting the rich creativity and diversity of local-scale practices around climate action we found during our research. We then conclude by focusing on the role and nature of quiet activism in the concluding chapter. Our research is particularly concerned with what causes practitioners and other actors who are living and working at the local scale to think and act the way they do in response to the climate emergency. This involves critical engagement with their framing of climate change, which serves to define problems, diagnose their causes, evaluate and make judgements about agents and impacts, and suggest remedies and predict their impacts.[39] This also involves identifying the kinds of learning that may (or may not) occur,[40] and how far they encourage (or inhibit) processes of creative discovery of new practices, through which new types of action, practices or frames can become recognised and adopted.

Chapter 2 *Building and Bridging the Knowledge Base* is focused on the first of five themes that underpin our understanding of quiet activism. This chapter explores how the shared understandings and collective capacity to address climate change are nurtured and developed at the local scale, and in doing so help to transform the status quo. At the heart of socially innovative responses to climate change are new knowledge, practices and connections between diverse people, groups and things (i.e. communities, neighbourhoods, sectors, nature, technology, art). Key to this is embracing the diverse nature of stories and recognising the rich possibilities of what counts as knowledge: Who can be a knower? (Scientists, politicians, citizens, animals/plants, children?); What is the purpose of knowledge? (To reinforce, educate, confuse, persuade, pacify, transform?); What kinds of things can be known? (Truth, beauty, right/wrong, past/present, future?); What are legitimate sources of knowledge? (Data, facts, stories, dreams, gossip?); and Who decides what knowledge is valid? (The government, community, experts, media?)

Chapter 3 turns to address the second key theme, *Bringing Missing Actors to the Table*. This chapter focuses on how responses to climate change have tended to focus on some actors and sectors to the exclusion of others. So how can we bring these missing actors to the table? Key questions include: what roles are different sectors playing (e.g. private sector, non-government organisations), and how can we activate missing actors? Central to this is the idea of *enabling* innovation. Innovation might be conceptualised as occurring across three domains—behaviour change, institutional change and built form change such as tree planting, lighter coloured roads and roofs, heatwave shelters, stormwater harvesting and so on. Elements of enablement include leadership, relationship building (e.g. trust, commitment, open communication, collaboration and information exchange), skills development and competence building (e.g. problem identification, developing solutions), decision-making support (e.g. financial or human resourcing, delegating power to act) and response-ability (i.e. to external disruptors, e.g. storm events or financial shocks, risk management, entrepreneurial readiness, experimentation).

Chapter 4 is focused on *Walking Together with Care*. This chapter explores the third key theme and the diverse collaborative practices of local climate action. Building community consensus around the urgency of climate change and the need for action around climate adaptation create a paradox of planning for certainty *and* planning for contingency. Under such circumstances those trying to implement climate action are pulled in many different directions, leading to fragmented rather than collaborative approaches which make it difficult to implement socially and environmentally innovative climate initiatives strategies. Walking together with care emphasises regenerative and creative practices that can create genuinely new opportunities for social and environmental change. Quiet activism involves rethinking local-scale action on climate change: the ways it *can* be done; the ways it *is* being done; and how this generative climate-related activity engages diverse and multiple collaborators at the local scale through "other" modes, narratives and activist practices.

The fourth key theme is outlined in Chap. 5 *Realising Transformative Potential*. This chapter focuses on the potential of local activists to expand their reach: to transform practices and social relations and empower actors more widely, more formally and/or more profoundly. This involves the need for significant changes to practices that are considered "normal". Such new practices are built on communities defined by shared interests, trust and solidarity. Without the support of such enabling relationships, many innovative ideas are only short lived or limited to small pockets of local activity. A second crucial aspect is empowerment as both a precondition of socially driven change and an important outcome of it; as participants in change-making practices gain collective resources (e.g. knowledge, social networks, infrastructures) which can be further mobilised to institutionalise, disseminate and/or deepen the impact of new practices. The interaction of these two processes—relationship building and empowerment—can lead to change beyond the ambit of the initial action.

In Chap. 6 *Making and Breaking Connections* the focus is on bringing together socially innovative activist practices (by local government authorities, non-government organisations and the private sector) and how they connect with other practices. In doing so we identify the critical points of connection and the potential footholds or leverage points for transformational change in response to the climate crisis. Vertical connections can be distinguished by three levels: *niches, socio-technical regimes* and *the broader socio-technical landscape.* Horizontal connections include those that exist between local interest groups, or across regimes. It is these points of connection which are important, as is having a multi-scalar perspective which considers links between niches, regimes and landscapes. These quiet activist practices can be creative disruptors, such as reading the landscape of legislation in new ways, or they can bring missing actors into conversations in order to engender transformative potential and politicised change to address the climate emergency.

The final Chap. 7 draws together the five key themes of this book by focusing specifically on the role of *Quiet Activism in Climate Change.* In doing so we return to questions around what it means to be an activist, or to "do activism" in the context of the climate emergency. Responding to climate change is a profound societal challenge and crisis that requires transformative change across society's institutions and structures. Our interest is at the local scale where we argue international, national and state policies are translated into practices, and community mobilisation, preparation and responses to the anticipated impacts of the climate crisis predominantly occur. Modest acts of care, connection and creativity can be collectively and politically significant. Through purposeful, collective commitment to socially innovative practices, local communities are forging new political pathways in response to the climate crisis. What is being described here is how working from everyday embodied practices with what people do and how they relate and make meaning; to a way of breaking down real and perceived barriers to addressing the climate emergency.

This we argue is the power and potential of quiet activism *in* climate change.

Notes

1. Flannery, T. (2020) *The Climate Cure: Solving the Climate Emergency in the Era of COVID-19,* Melbourne, The Text Publishing Company.
2. See Baker, S. and Mehmood, A. (2015). Social innovation and the governance of sustainable places, in *Local Environment: The International Journal of Justice and Sustainability* 20(3), pp. 321–334.
3. See Chatterton, P. and Pickerill, J. (2010) Everyday activism and transitions towards post-capitalist worlds, in *Transactions of the Institute of*

British Geographers, 35, pp. 475–490; Hackney, F. (2013) Quiet activism and the new amateur, in the Journal of Design and Culture, 5(2), pp. 169–184; Pottinger, L (2017) Planting the seeds of quiet activism, in Area, 49(2), pp. 215–222.
 4. See Eisenmann, L. (2005) A time of quiet activism: Research practice, policy in American women's higher education 1945-1965, *History of Education Quarterly* 45(1), pp. 1–17; Clarke, K (2016) Willful knitting? Contemporary Australian craftivism and feminist histories, *Continuum*, 30(3), pp. 298–306.
 5. See for example Tactical Urbanist's Guide to getting it Done, accessed on http://tacticalurbanismguide.com/about/
 6. Rose, D. B. (2004) *Reports from a Wild Country: Ethics for Decolonization*, University of New South Wales Press, Sydney.
 7. Ibid.
 8. Steele, W. and Rickards, L. (2021) *The Sustainable Development Goals and Higher Education: A Transformative Agenda?*, London, Palgrave
 9. Rose, D.B (2004).
10. Ibid., p. 18.
11. Ireland, P., & McKinnon, K. (2013). Strategic localism for an uncertain world: A postdevelopment approach to climate change adaptation. *Geoforum*, *47*, 158–166.
12. Steele, S., MacCallum, D., Byrne, J. and Houston, D. (2012) Planning the Climate-just City, in *International Planning Studies*, 17(1), pp. 67–83.
13. Rauken et al., 2014.
14. See Byrne, J., Gleeson, B., Howes, M., Steele, W. (2009), Climate change and urban resilience: the limits of ecological modernization as an adaptive strategy, in Davoudi, S, Crawford, J and Mehmood, A (Eds) Planning for climate change: Strategies for mitigation and adaptation for spatial planners, Earthscan, London, pp. 136–54; MacCallumD; Byrne, J Steele, W (2014) Whither Justice? An Analysis of Local Climate Change Responses from South East Queensland, Australia. *Environment and Planning C: Government and Policy* 32:1, pages 70–92.
15. Dupuis, J and Knoepfel, P (2013), The adaptation policy paradox: the implementation deficit of policies framed as climate change adaptation, Ecology and Society, 18 (4), p. 31.
16. Moulaert, F., D. MacCallum and J. Hillier (2013) Social Innovation: intuition, precept, concept, theory and practice, in F. Moulaert et al. (eds.) *International Handbook of Social Innovation*, Elgar, Cheltenham.
17. Scott-Cato, M. and J. Hillier (2010) How could we study climate-related social innovation? Applying Deleuzean philosophy to Transition Towns. *Environmental Politics* 19(6), 869–87.
18. Deleuze G. and F. Guattari (1987) *A Thousand Plateaus: capitalism and schizophrenia,* Athlone Press, London.

19. See for example Marques et al. (2017) for quantitative analysis of the term's appearance in published work.

20. Many recent literature reviews identify distinct and possibly incommensurable (Montgomery 2016) schools of thought in relation to social innovation (e.g. Montgomery 2016; Unger 2015; Ayob et al. 2017; Jessop et al. 2013; Nichols et al. 2015; Shockley 2015; van der Have and Rubalcaba 2017). Our position, as outlined here, is aligned with the perspective often associated with urban and regional studies, rather than with business and management (Moulaert et al. 2013).

21. Moulaert, F., D. MacCallum, A. Mehmood and A. Hamdouch (2013) General introduction: the return of social innovation as a scientific concept and a social practice, in F. Moulaert et al. (eds) *International Handbook of Social Innovation*, Elgar, Cheltenham.

22. Ibid.

23. See Howaldt, J. and M. Schwarz (2016) 'Verifying existing Social Theories in reference to Social Innovation and its Relationship to Social Change', *SI-DRIVE Deliverable* 1.3; Garcia, M., and S. Vicari Haddock (2016). Special issue: housing and community needs and social innovation responses in times of crisis. *Journal of Housing and the Built Environment* 31(3), 393–407.

24. Moulaert, F., D. MacCallum, A. Mehmood and A. Hamdouch (eds.) (2013). *International Handbook of Social Innovation: collective action, social learning and transdisciplinary research*, Elgar, Cheltenham.

25. Intergovernmental Panel on Climate Change (IPCC), (2013) Climate Change 2013: The Physical Science basis, accessed on https://www.ipcc.ch/report/ar5/wg1/

26. Schultz, J. (2011) Life in a Time of Disasters, in *Surviving*, Griffith Review, 35, Griffith University, pp. 7–11.

27. O'Malley, N. (2020) *World awaits action by 'suicidal' Australia, says former climate chief*, in The Age, December 1st, accessed on https://www.theage.com.au/environment/climate-change/world-awaits-action-by-suicidal-australia-says-former-climate-chief-20201201-p56joj.html

28. Hao Z., Singh V.P. and Hao F. (2017) Compound Extremes in Hydroclimatology: A Review, *Water*, 10(6): 1–24 / Kopp RE, Hayhoe K, Easterling DR, Hall T, Horton R, Kunkel KE, and Kopp, R. E., Hayhoe, K., Easterling, D. R., Hall, T., Horton, R., Kunkel, K.E., & LeGrande, A. N. (2017). Potential surprises—compound extremes and tipping elements. In D. J. Wuebbles, D. W. Fahey, K. A. Hibbard, D. J. Dokken, B. C. Stewart, & T. K. Maycock (Eds.), *Climate science special report: Fourth national climate assessment, Volume I* (pp. 411–429). U.S. Global Change Research Program.

29. Steffen, W., Dean, A. and Rice, M. (2019) *Weather Gone Wild: Climate Change fuelled extreme weather*, Climate Council of Australia, accessed on

https://www.climatecouncil.org.au/wp-content/uploads/2019/02/
Climate-council-extreme-weather-report.pdf

30. SeedMob: https://seedcrowdfunder.raisely.com/aboutseed
31. Birch, T. (2017) *Climate Change, Recognition and Social Place-Making*. Sydney Review of Books. Accessed on https://sydneyreviewofbooks. com/essay/climate-change-recognition-and-caring-for-country/
32. Sandercock, L and Forsyth, A (1992) A Gender Agenda: New Directions for Planning Theory, *Journal of the American Planning Association*, 58(1):49–59.
33. Belenky, M, Clinchy, B, Goldberger, N, Tarule, J. (1986) *Women's ways of knowing: the development of self, voice and mind*, NY: Basic Books.
34. See for example Burch, S. (2010), Transforming barriers into enablers of action on climate change: Insights from three municipal case studies in British Columbia, Canada, *Global Environmental Change*, 20, 287–297; Neumeier, S. (2017), Social innovation in rural development: identifying the key factors of success, *The Geographical Journal*, 183(1), 33–46; Scott-Cato, M. and Hillier, J. (2010), How could we study climate-related social innovation? Applying Deleuzian philosophy to Transition Towns, *Environmental Politics*, 19(6), pp. 869–887.
35. Pasquini, L., Ziervogel, G., Cowling, R.M. and Shearing, C. (2015) What enables local governments to mainstream climate change adaptation? Lessons learned from two municipal case studies in the Western Cape, South Africa, *Climate and Development*, 7(1), 60–70.
36. See Whyte, K.P. (2017) Our Ancestors Dystopia Now: Indigenous conservation and the Anthropocene. In Ursula K Heise, Jon Christensen and Michelle Neimann (eds) *The Routledge Companion to the Environmental Humanities*. London: Routledge.
37. See Rip, A. & Kemp, R. (1998). Technological change. In Rayner, S. & Malone, E. (Eds.), *Human Choice and Climate Change: Resources and Technology*, Vol. 2. Columbus, OH: Battelle Press; Geels, F. (2011). The multi-level perspective on sustainability transitions: responses to eight criticisms. *Environmental Innovation and Societal Transitions*, 1, 24–40; Hargreaves, T., Longhurst, N. & Seyfang, G. (2013). Up, down, round and round: connecting regimes and practices in innovation for sustainability. *Environment & Planning A*, 45, 402–420.
38. Shove, E. (2003). *Comfort, Cleanliness and Convenience: the Social Organization of Normality*. Oxford: Berg.
39. See Rein, M and Schon, D (1994) Frame Reflection: Toward the resolution of intractable policy controversies, Basic Books, New York.
40. See Healey, P (2008) *Urban complexity and spatial strategies: towards a relational planning for our times*, Routledge, London.

CHAPTER 2

Building and Bridging the Knowledge Base

Abstract This chapter explores how the shared understandings and collective capacity to address climate change are nurtured and developed at the local scale, and in doing so help to transform the status quo. Key to this is embracing the diverse nature of stories and recognising the rich possibilities of what counts as knowledge. At a local scale this range of ways of "knowing" involves a spectrum of stories: informal, subjective accounts such as dialogue, personal experience, local knowledge, visual images, actions and activities; as well as the more formal, objective reports such as resource mapping and scientific research. Building and bridging the knowledge base involves engaging with diverse ways of knowing through a constellation of stories, experience and activities—the innovative activist practices and relationships that take root and flourish at the local scale.

Keywords Climate change • Knowledge • Stories • Informal • Activism • Practices

© The Author(s), under exclusive license to Springer Nature
Switzerland AG 2021
W. Steele et al., *Quiet Activism*,
https://doi.org/10.1007/978-3-030-78727-1_2

CLIMATE ACTION HEARTLANDS

At the heart of quiet activism and socially innovative responses to climate change at the local scale are the knowledge and practices that help develop the shared understandings and collective capacity needed to transform the status quo. This chapter focuses on the first of the five key themes in the quiet activist framework. Building and bridging the knowledge base involves making new connections both in terms of knowledge, skills and practices, but also between diverse people, groups and things (i.e. communities, neighbourhoods, sectors, nature, technology, art). We will return to the importance of making and breaking connections in Chap. 6.

This is about the power of stories. Key to this is embracing a wide range of different (sometimes conflicting) narratives and storied practices, as well as recognising the rich diversity of what counts as knowledge and who counts as a knower when taking action in a climate of change. There is a rich "multiplicity of ways of knowing"[1] or "connected knowing"[2] underpinned by important, yet oft overlooked or assumed questions such as:

- Who can be a knower? (Scientists, politicians, citizens, animals/plants, children?)
- What is the purpose of knowledge? (To reinforce, educate, persuade, transform?)
- What kinds of things can be known? (Truth, beauty, right/wrong, past/present, future?)
- What are legitimate sources of knowledge? (Data, facts, stories, dreams, gossip?)
- Who decides what knowledge is valid? (The government, elders, community, experts, media?)

This range of ways of "knowing" the climate emergency at the local scale includes a spectrum of stories: informal, subjective accounts such as dialogue, personal experience, local knowledge, visual images, actions and activities; as well as the more formal, objective reports such as resource mapping and scientific research. Stories in all their forms are the compasses by which we navigate our place in the world, offering a way of travelling from here to there and locating the silences in between.[3] Stories help build knowledge through the narratives, ideas, anecdotes, summaries, histories, musings, theories and data that work to create empathy and identity. Place itself for example is a set of local story/storied/stories.

The everyday lived practices and experiences of climate change at the local scale are a storied locale or landscape. When Indigenous Australian award-winning artist and Gunditjmara/Bundjalung man Archie Roach, for example, sings *Go Down to the River*, it is a heartfelt, moving ballad. One that speaks through—and to—his life-long connections with his now deceased partner Ruby Hunter, a Ngarrindjeri, Kokatha Nga and Pitjantjatjara woman who was born on the banks of the Murray River.[4] In this song/story he sings of the beauty of life on the River and its people, of the Dreamtime and the current state of climate change and his hopes for the future of the Ngarrindjeri culture, which are all intimately intertwined.

This is linked but different to the story by journalist Margaret Simons *Cry me a River* in which she evocatively describes the urban food bowl of Eastern Australian, the Murray–Darling River Basin, as: 'like a tree, except the sap runs not from root to twigs but in the other direction. Here the river runs pea-green between red cliffs through semi-desert country to the vast sheets of water that form the lower lakes, the Coorong and the sea'.[5]

Quiet activism at the local scale involves a constellation of stories, experience and activities that link together formal and informal practices identified in Fig. 2.1. These stories help to build the localised knowledge and shared connections that lead to empowerment. Socially innovative climate action are stories about practices: ways of becoming and knowing at the local scale. The story of climate action as quiet activism is where alternative pathways and futures can be critically forged through a mix of these different types of stories. The spaces, places and practices are where we seek out new stories that are struggling to be born from the "compost of the old".[6]

Stories help to build and refine the shared knowledge and practices that are needed in a transformative approach to climate change. Through sharing local stories across all sectors (public, private, not-for-profit), the dominance of technical and institutional "fixes" of climate adaptation gives way to an ethics of connection.[7] Quiet activism in this sense represents different knowledge and practices that may seem disconnected but constellate to frame stories and spaces of climate-justice at the local scale. This involves identifying and developing counter-hegemonic praxes that enable us to re-imagine our experience of, and responses to, climate change.[8] As Bruno Latour notes:

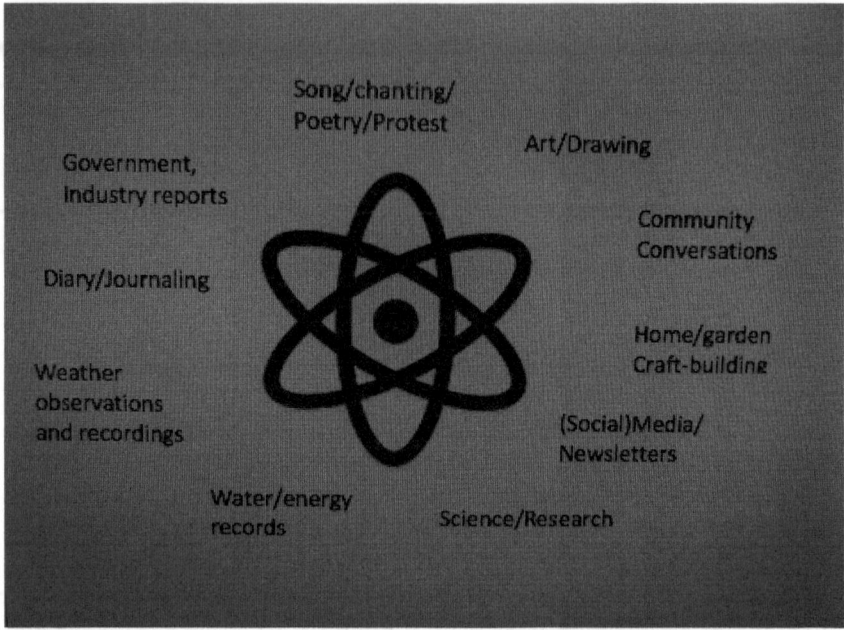

Fig. 2.1 A constellation of stories for local climate change action. (Source: Authors)

> A common world is not something we come to recognize, as though it had always been here (and we had not until now noticed it). A common world, if there is going to be one, is something we will have to build, tooth and nail, together.[9]

This in turn resonates with Isabelle Stenger's idea of cosmopolitics[10]—where diverse stories, perspectives, experiences and practices can connect to create the foundation for new strategic possibilities. Our approach is empirically informed by conversations with diverse actors at the local scale who are mobilising different approaches to this ideal in various grassroots actions on climate change. What follows are four very different stories that reveal different ways of building and bridging the knowledge base in the purposeful search for transformative change in response to the climate emergency at the local scale.

A Local Mandate for Climate Action

Climate for Change is a not-for-profit organisation with a mission to create the social environment needed for effective action on climate change. The organisation arose in 2014 in Melbourne, Australia, in recognition that inadequate political action was reflective of limited engagement with Australians outside the "climate change bubble". Rather than seeking to "preach to the converted", Climate for Change has been working hard to embed climate change into the mainstream public consciousness. By working at the individual and household level, they have engaged thousands of Australians not through public lectures or rallies—but through kitchen table style conversations with family and friends.

To achieve their core mission to create a social mandate for effective government action, Climate for Change engages everyday people in democracy through peer-to-peer discussions on climate change. They do this by working within networks of trust to reframe climate action as a social norm, thereby enabling people to face up to and engage with a potentially contentious issue. As such, Climate for Change is bridging the gaps between scientific knowledge, citizen engagement and political ambition. Their stated mission is to create the social climate for effective action on climate change.

> We know that social change happens when ordinary people start having conversations with those they know and trust. We support people to have effective conversations about climate change, and to take actions that will inspire others.[11]

Climate for Change is based around facilitating local-scale conversations around the kitchen table about the drivers of climate change and the need for action. Under this model, an individual nominates to host a gathering at their house. At this gathering, attendees are encouraged to host subsequent events with their own social networks—thereby growing Climate for Change's outreach. The organisation uses this model to access people who wouldn't normally engage with the issue of climate change in a proactive manner, such as by attending a public rally or lecture.

Key to Climate for Change's work is engaging people on their own terms. To do this, the trained facilitators create an inclusive and open forum where people are encouraged to ask questions and have the time and space to consider how climate change impacts them personally. The conversations are based on existing social networks, with the topic of climate change and a trained Climate for Change facilitator being the only added variables. As such, climate change is brought into a comfortable and familiar setting to attendees where every aspect of the event is "normalised", except for the topic of conversation.

> We ... specifically focussed on helping people to have conversations with their peers on climate change – something that is now being recognised by experts as key to building public support for the action we need on climate change.[12]

Climate for Change frames their work as a democratic project in citizen education and participatory climate action. Attendees are asked to reflect on their own beliefs and values through conversation and are provided with specific ideas and actions they can pursue after the event. The group encourages people to go beyond individualistic action to amplify their impact, such as by hosting a subsequent gathering or writing to their Member of Parliament. The organisation asks people to reflect on whether climate change is important to them as an individual, and why this is the case.

> Purposeful conversations are a great opportunity to shift general concern about climate change into effective and meaningful action ... You don't need to be an expert to talk to others about climate change - just a person who cares deeply about our future!... Powerful conversations are all based on human connection. We connect by asking thoughtful questions, listening to understand, being authentic, and sharing something of ourselves in these conversations. We build hope and confidence by giving people a sense of what is possible and what is required and showing that the things they care about fit into a positive shared vision for the future.[13]

Through its focus on helping everyday people have peer-to-peer conversations about climate change, Climate for Change occupies a unique and important space within the NGO sector. Climate for Change hopes their conversations are a catalyst for climate action, particularly for attendees who wouldn't normally engage with the issue (Fig. 2.2). In this way, an objective

Fig. 2.2 Actively creating a climate for change. (Source: https://www.climate-forchange.org.au/about)

of the organisation's work is to increase public education and civic engagement. This contrasts with other public conversation forums which often have a specific outcome in mind for the organiser—such as focus group consultations held by local governments to feed into council strategies.

At each Climate for Change event, the facilitator begins by showing a video which illustrates formal and objective ways of knowing climate change, including the science behind the greenhouse effect and feedback loops. This video is used as a stimulus for discussion with attendees, as the facilitator encourages people to move beyond objective reasoning to connect with climate change on an emotional level. Facilitators connect the objective with the emotional by using science as a starting point, asking questions like: *How does this knowledge make you feel?* Climate for Change is bridging the gap between objective/formal and subjective/informal ways of knowing climate change through meaningful and purposeful local-scale conversations.

In addition to *bridging* the gap between knowledge and action, Climate for Change is also *building* new knowledge on how to have effective and meaningful conversations about climate change. Research to date has tended to focus on mass communication and the psychology of climate change, rather than the importance of peer-to-peer conversations. Climate for Change has found that two aspects are particularly important for effective peer-to-peer conversations—being a good listener and asking the right questions.

What is most important is the conversations which follow and the avenues that are provided for ongoing engagement and involvement. The science and objective knowledge presented in the video is seen as an enabler, rather than an end point. At the end of the event, facilitators provide options for people wishing to act, such as hosting a subsequent Climate for Change event, joining the facilitator community or writing to local members of parliament. In this way, Climate for Change is also bridging the gap between education about climate change and action on climate change. The aim is to amplify 'the power of everyday people to shape democracy when they actively engage as citizens'.[14]

To help attendees engage with their local members, Climate for Change is currently supporting local "MP Engagement Groups" run by other volunteers in their community. These events provide people with the opportunity to write to their MP in a collaborative and supported space. Climate for Change aims to influence federal climate policy and discourse through active engagement with the local community (see Fig. 2.2). Events tap into existing social networks, which are inherently place-based and localised in nature. Attendees are encouraged to engage with their local member for parliament, for example, to stress the urgency of the climate crisis and to demand better policy as Sue's story of being a citizen climate activist below illustrates.

I'm Sue. Teacher, writer, parent, and passionate advocate for government action on climate. I'm also a volunteer facilitator with Climate for Change and the MP Engagement Groups coordinator. By late 2017, I'd been a facilitator for about a year. I'd talked to many people about the power of citizen democracy but had never visited an MP myself. So, on a whim, deciding I should "walk the talk" I turned up nervously to a Preston cafe in response to a request for volunteers to visit my then local MP, David Feeney. Unwittingly, this single action would lead me to create a project supporting many community groups to engage with their MPs on a regular basis. A dozen locals turned up for the briefing … David gave in to our pressure. He signed a short statement opposing coal and had a group photo taken with him holding a Stop Adani[15] sticker. He agreed to post it on social media and also to speak in Parliament about coal and climate. The Guardian saw his tweet and ran a story. David spoke in Parliament and the Guardian ran a second story. I was amazed. This was proof citizen democracy could work.[16]

The theory of change central to Climate for Change's work is that the more people they can reach through local-scale gatherings, the more engagement there will be between citizens and government. This can then be the base for more ambitious and urgent government action on climate change. Facilitators attend gatherings to generate and steer discussion amongst attendees. Rather than being climate change "experts", facilitators are everyday people who have passion and drive for the issue and have been trained to hold (at times) emotionally demanding conversations. *Knowledge of climate change is good, but only if you know how to act,* says Blanche Verlie, a trained Climate for Change facilitator. According to Verlie, when it comes to action on climate change at the local scale,

We don't have a knowledge deficit; only an empowerment deficit.[17]

THINK + ACT + SHARE = CHANGE

Green Cross Australia is a leading not-for-profit organisation that educates and empowers everyday citizens to adapt to their changing environments at the local scale. The organisation fills the space between humanitarian and environmental challenges (where there was previously a void in the Australian climate action landscape). The not-for-profit organisation embraces the idea that existing challenges can be reframed to present opportunities for a better, more resilient future by purposefully building and bridging the knowledge base with local citizens, businesses, government agencies, community and research partners.

Green Cross Australia empowers a resilient Australia. We educate and empower people and businesses to become more resilient to our changing climate and environmental stresses. Together with our partners, we run innovative, digital projects, which help to grow more sustainable communities. We are helping our cities and coasts to prosper by advancing food security, property resilience, and urban landscapes.[18]

Since its inception, the organisation has connected with people who feel "put off" or paralysed by politics, scare tactics and complex science through a resilience lens, rather than a climate change or sustainability paradigm. By deliberately framing their work around resilience—particularly building social inclusion and the ability of communities to bounce back from stresses—Green Cross Australia has been able to gain more traction and buy-in for their work. Steering clear of using "climate change"—a polarising term in Australia—to describe their work has helped build and bridge a range of community-based activities and conversations. Examples can be found in a number of their projects and programmes including the Business Adaptation Network, Harden Up—Protecting Queensland, Witness King Tides, Build it Back Green, Green Lane Diary, Future Sparks and the Queensland Climate Adaptation Strategy.

Green Cross Australia is a member of Green Cross International—a global organisation founded in 1993 by Nobel Peace Prize laureate and former president of the USSR, Mikhail Gorbachev, following the Rio Earth Summit. Green Cross Australia similarly emerged in response to shifting climate change narratives and on the back of federal political change in Australia, including the signing of the Kyoto Protocol by former Prime Minister Kevin Rudd. Much like the traditional humanitarian and emergency response organisation Red Cross, Green Cross is not a political advocacy group. Rather than lobbying governments to implement policies and programmes to build the resilience of Australians, Green Cross Australia is taking charge of the change.

> We are not an advocacy group. We work alongside respected business, research community and government partners to deliver world-class digital values and behaviour changing projects. We strive for a secure and sustainable future.[19]

Green Cross initiatives seek to empower everyday Australians to take practical, informed action and provide them with social and digital media tools to share their actions with friends. Green Cross Australia focuses on the use of socially innovative digital projects to help enable more sustainable and resilient communities. The organisation embeds mapping technologies in its digital platforms to visualise participation in its projects and measure engagement. Through their website and digital platforms, Green Cross Australia have engage Australians in locally significant projects. They work to foster local networks and partnerships through digital communications, creating the social fabric necessary for community resilience. This is outlined by Jeremy Mansfield, Deputy Chair of the Green Cross Australia Board.

Each of us has to develop this very deep well of self-reliance, but in order to do that, we have to be connected to the social fabric around us. That is the conundrum of how we get through a change in climate – we have to take personal responsibility and develop the capacities to adapt and be vulnerable, and understand shock and mourn, but at the same time be optimistic and positive and prepared, and there for each other.[20]

The *Business Adaptation Network* (BAN) for example is Australia's only climate change and resilience network focussed on business. BAN members include organisations from the private sector such as Australia Post, Optus, Ernst & Young and Suncorp, as well as members from government and research organisations. Being part of BAN provides the opportunity to collaborate with a network of like-minded professionals to develop real-world business and community solutions and responses to climate change. It also provides the opportunity to participate in a range of events designed to keep up to date with the latest climate change and resilience research and knowledge.

The BAN helps to foster business engagement, and the private sector funding from its supporters has helped to support and scale-up activities including: leadership and industry collaboration; creative engagement that enables capacity building and cross-sector opportunities; networking opportunities to create connections and partnerships; and joint engagement on climate risk and adaptation scenarios at the local scale.

> Business engagement with climate resilience is critical to building socio-economic resilience and has the ability to innovate, develop strategies and opportunities on resilience, and cater to new market needs as people adapt to new conditions. Business ingenuity and resources are needed alongside government efforts and community support to address climate change and its consequences.[21]

Green Cross Australia facilitates and fosters face-to-face conversations across business, government, community and academic stakeholders. They do this by tapping into existing and established networks and forums where possible, rather than recreating the wheel. Like many NGOs, Green Cross Australia has been affected by funding constraints. However, they continue to be motivated to share their skills and expertise to build and bridge the knowledge base needed for local action on climate change.

Several initiatives serve as knowledge hubs for local communities. These include:

- *Build it Back Green*—which assists disaster affected communities to build back sustainably through the co-development of tools and resources.
- *Green Lane Diary*—works with school children to identify ways to take everyday actions in their community that can make a difference in climate change.
- *Witness King Tides*—creatively engages community arts to focus on amplifying the effects of climate change on oceans and coastal communities.
- *Extreme Weather Heroes*—mobilises young people within local communities to become emergency volunteers, including the use of social networking for climate action.

The Green Cross mantra—*Think + Act + Share = Change*—is central to their work and the "building and bridging knowledge base" theme. Recognising that governments alone will not be able to deal with the scale of climate change–induced impacts, Green Cross has sought to build capacity, share knowledge and encourage practical action. Doing this has meant recognising that climate change means very different things to different people in the community, and bridging skills and knowledge around action on climate change is most effectively communicated and practised at the local scale. They seek to be thinkers, connecters and doers sitting at the nexus of a humanitarian organisation and an environmental organisation, with an emphasis on community action on climate change and building the knowledge and capacity for local responses to this.

LOCAL COMMUNITY CLIMATE HUB

Established in early 2001, *Environment House* is a volunteer-led organisation in Bayswater, Western Australia, focused on sharing resources and supporting a shift to a more sustainable way of life for the wider community. The aim is 'to both "walk the talk" and "talk the walk" on climate protection, river and wetland care and other sustainability issues'.[22] This is reflected in their constitution which commits them to take action on climate change at the local scale through community education and involvement. As the organisation describes in their online backstory:

Late in 2000, three friends (Brenda Conochie, Rob Gulley and Rachael Roberts) decided a shopfront community environment centre was needed in Perth – a campaign centre and eco-products store to support the community's efforts to live in a way that respects our environment and future generations. We gathered others to form a community association for this purpose. With underwriting by the ACF (Australian Conservation Foundation), we leased a former fish & chip shop plus house and backyard, in the heart of Maylands, only 150m from the train station, and 8 minutes from the city. Generous well-wishers helped with the private commercial rent, and the first furnishings and office equipment. We had a big launch party in March 2001, with guests of honour Peter Garrett (then Midnight Oil front man), Judy Edwards (then Minister for the Environment) and a Noongar elder who has since passed away.[23]

As a hub for eco-knowledge, the volunteers at Environment House seek to share their practices, experiences and stories through workshops and events to build the capacity of the community. The workshops feature waste management practices, organic gardening techniques, and household energy and water efficiency measures in direct response to the local context and needs of the community. Workshops are run throughout the school holidays in conjunction with local schools and childcare centres. These workshops spark curiosity in the local youth and encourage changes in the way they perceive and interact with the environment and sustainable practices.

Other events include interactive tours of local green spaces and wetlands to help residents discover and connect more deeply with their local environments through storytelling and conversation. Regular *Share Your Spare* events also encourage the sharing of excess home grown produce and seeds, fostering relationship building and knowledge sharing amongst community members. Local activities are co-hosted with other community groups such as *Transition Towns Bayswater*[24] which offers an open invitation for the community to "share and repair" on the first Saturday of every month.

Environment House operates as a local hub for the community, offering an inclusive space where individuals can seek advice, learn skills and engage in a community actively and collaboratively adapting to climate change at the local scale. The eco-shop for example facilitates conversations and sharing of ideas that lead to tangible changes in consumption practices. The shop has supported local eco-businesses to grow and thrive by providing a physical space and platform for the marketing of sustainable goods and services.

Taking a partnership approach, Environment House has worked closely and effectively with local households, schools, aged care facilities, child-care centres, migrant centres and various local and state government authorities to provide energy and water efficiency audits to different segments of their community. Working with Hillcrest Primary School, for example, Environment House conducted an energy and water efficiency audit, enabling the implementation of new practices. In addition, these audits presented an opportunity to educate and engage students and staff on sustainability. This helped to bridge knowledge gaps and build the capacity for transformative change within the school and in the individual homes of students and staff.

Environment House working with the City of Bayswater also developed and delivered the Bayswater $WAP Program in 2015. Under this grassroots initiative, over $11,000 worth of eco-hardware was given away to local residents to reduce household energy and water use. Through the Bayswater $WAP Program, Environment House provided expertise to build the knowledge of participating households. Workshops and conversations with participants provided information—such as how to select and install the right appliances for the right household—ensuring sustained outcomes from the programme.

The team behind Environment House were awarded funding to develop a *Cool Communities* demonstration house. With this funding, the team established the centre as a gateway for knowledge sharing on sustainability and adaptive climate change practices. By utilising the expertise of their trained sustainability auditors and building strong and lasting relationships with government agencies, the organisation has been able to undertake resource consumption audits and implement cost-saving measures in individual households, schools and businesses across the greater Perth region. The organisation has progressively evolved by taking funding opportunities and initiating projects that have enabled them to extend their reach and share their collective knowledge. This in turn has built the capacity of individuals and organisations in Perth to live and operate more sustainably.

Through these programmes and services, Environment House highlights how the stories and knowledge generated through their work support building and bridging knowledge at the local scale. This has helped residents, schools and businesses better understand and realise savings in their energy and water usage. The constellation of experiences and knowledge offered by the volunteers at Environment House provides the community with the expertise needed to evaluate and challenge the status quo. This has catalysed transformative changes to the individual practices and understandings of climate change held by members of the Bayswater community (and beyond).

> From seed we collected along the river, we grew thousands of she-oaks, paperbarks and rushes, and planted them out on the eroded Maylands foreshore. We composted tonnes of scraps from the Fruit & Veg just down the lane. We've taught composting, worm farming and bokashi to hundreds of folks, we hosted teaching students; and gave talks to hundreds of community groups on sustainable living.[25]

ReNew Initiatives

Our final story for this theme of quiet activism on building and bridging the knowledge base focuses on the *ReNew* Initiatives which demonstrate technology in action and provide practical knowledge for people wishing to adopt more sustainable lifestyles. The *ReNew* magazine is a quarterly publication produced by the Alternative Technology Association that showcases practical sustainable living and innovations in sustainable building practice and community-based renewable energy technology. This provides a stepping-stone for local climate change initiatives by building awareness of local-scale solutions that are accessible to individuals and community groups.

The Pears Report[26] is a regular column written by Alan Pears for the *ReNew* magazine. A collection of Pears' best columns was published as a collection by the Alternative Technology Association. Each of Pears' columns appears alongside a series of articles reviewing evolving perspectives on key themes, such as energy efficiency, efficient appliances (stars, schemes and standards), building and energy policy, energy market reform and climate policy. Each column is backed up by formal/objective forms of knowledge such as scientific research and data.

As a regular contributor and climate knowledge builder and broker, Pears has written a column for every edition of *ReNew* since 1997. He has used the column as a bridge to examine the complexities and absurdities of climate and energy policy in Australia. Pears' work has challenged the dominant political discourse and the current energy system by highlighting sustainable alternatives. His work has explored energy efficiency of buildings, environmental ratings and regulation, green building and appliance innovation, and commercial and industrial energy management.

The Pears Report: Reflections on two decades of energy and climate policy celebrates building and bridging the Australian knowledge and understanding of climate mitigation and adaptation at the local scale. Pears' recent columns have addressed topics like the fragility of Australia's traditional energy grid in times of peak energy demand, as well as the feasibility and trend towards diversified and distributed energy systems. He has also recently explored the difficulties faced by vulnerable households in accessing sustainable energy options, highlighting the options for local communities and advocating for ways to overcome this as a climate justice imperative.

> In the past, financial institutions loaned money to governments or big businesses to build power stations and gas supply systems. Now we need mechanisms to give all households and businesses access to loans to fund the new energy system. Households that cannot meet commercial borrowing criteria, or are disempowered – such as tenants, those under financial stress, or those who are disengaged for other reasons – need help.[27]

THE QUIET STORIES OF CHANGE

Climate change sometimes figures as a site in which long-standing debates are re-enacted and one in which the scale and character of what constitutes climate activism calls for new ways of thinking at different scales. As "ecologies of practices", climate activism at the local scale focuses on the modes and products of action in ways that matter, in ways that shape the future and thus represent a way of disentangling local-scale action on climate change from the apathy and/or cynicism that can infect more formal arenas of politics and policy. Following Isabelle Stenger's, quiet activism at the local scale is an invitation to: 'think in the presence of – as this profoundly reshapes the nature and types of questions that we ask of the present in ways that leave no-one unaffected'.[28]

Building and bridging the knowledge base through localised stories and socially innovative practices is a key component of climate action and the climate-just city and region. One that highlights the very particular and contextualised ways that place, history, culture are entangled with questions of justice and the need for, and nature of, local-scale activism, as an enlarged concept 'encompassing relational, emotional, embodied place-based knowledge and practices'.[29] The power of the stories highlighted above and their associated local practices serve two key functions in justice struggles that are both "inventional" and help to create new meanings that work to test and challenge the status quo.

> first to set the stage for the exploration of potential futures, and second to enact new possibilities. That is stories are *inventional*: they reassemble the chronology and causality of past events to produce *new meanings* capable of challenging the unacknowledged narratives capable of suppressing the unacknowledged narratives that sustain oppressive conditions.[30]

Quiet activism as a focus on localised practices of "making and doing" distinguishes between working with *detached* knowledge and working with *locally relevant* knowledge that seeks to build and bridge the knowledge base in diverse ways. This does not mean rejecting meta-science or policy, but it does call into question what they actually do, and how they might be put better to work within the context of local places, spaces and communities as a means of developing practical, situational, regenerative politics in the context of the climate emergency. This in turn reveals the emergence of new forms of practical interactions which provide capacity for how climate activism and action might be re-imagined by communities as localised practices of kinship and care.

In this chapter building and bridging the knowledge base reflects a range of different contexts as well as material and discursive realities that serve as a political space where individuals and communities come together to interpret and act on the climate emergency. In each of the examples outlined in this chapter—Climate for Change, Green Cross Australia, Environment House and the *ReNew* Initiatives—the emphasis was on identifying and implementing socially innovative responses for localised climate action. These are the stories and practices of quiet activism that seek to create change at the local scale. In the following chapter we turn to focus on the second key theme from the quiet activist framework. This is about enabling innovative practices as a form of societal and civic response-ability by bringing missing actors to the climate action table.

Notes

1. Sandercock, L and Forsyth, A. (1992) A Gender Agenda: New Directions for Planning Theory, *Journal of the American Planning Association*, 58(1):49–59

2. Belenky, M, Clinchy, B, Goldberger, N, Tarule, J. (1986) *Women's ways of knowing: the development of self, voice and mind*, NY: Basic Books.

3. Solnit, R. (2013) *The Faraway Nearby*, New York: Viking.

4. See   https://www.theage.com.au/entertainment/art-and-design/kura-tungar-20041019-gdytpk.html

5. Simons, M (2020) *Cry Me a River*, The Quarterly Essay, Issue, 77, Melbourne, Black Inc. Publishing.

6. Mueller, M (2017) *Being Salmon, Being Human: Encountering the Wild in Us and Us in the Wild*, White River Junction, Chelsea Green Publishing.

7. Rose, D and Robin. L. (2004) The Ecological Humanities: An Invitation. *Australian Humanities Review*, 31–32, accessed on http://australianhumanitiesreview.org/2004/04/01/the-ecological-humanities-in-action-an-invitation/

8. Houston, D., MacCallum, D., Steele, W., Byrne, J. (2016) Climate Cosmopolitics and the possibilities for urban planning, *Nature + Culture*, 11(3): 1–29

9. Latour, B. (2010) An Attempt at a Compositionist Manifesto. *New Literary History* 41(3): 471–490.

10. Stengers, I (2010) *Cosmopolitics I, II*. Trans. Bononno, R Minneapolis: University of Minnesota Press.

11. Climate for Change accessed on https://www.climateforchange.org.au/

12. Ibid.

13. See—*The Conversation Guide*, accessed on https://drive.google.com/file/d/1Tt6-ueh8vdS0Jf3WkGK4Jr3OEduGANxA/view

14. Climate for Change accessed on https://www.climateforchange.org.au/

15. "Stop Adani" refers to a grassroots movement to stop the Adani coal mine in Queensland, see for example https://www.stopadani.com/

16. Available   on   https://www.climateforchange.org.au/becoming_a_citizen_climate_activist

17. Quotation taken from Australian Research Council (ARC) DP150100299 individual interview transcripts

18. About Green Cross Australia accessed on https://www.greencrossaustralia.org/about-us.aspx

19. Ibid.

20. Quotation taken from Australian Research Council (ARC) DP150100299 individual interview transcripts.

21. Ibid.

22. Environment House backstory accessed on https://www.envirohouse. org.au/view/about
23. Ibid.
24. Transition Towns Bayswater see—https://transitionaustralia.net/group/ transition-town-bayswater/#:~:text=Transition%20Town%20 Bayswater%20(TTB)%20is,and%20resilient%20way%20of%20life
25. Environment House backstory accessed on https://www.envirohouse. org.au/view/about
26. See—*The Pears report: reflections on two decades of climate change and energy policy* accessed on https://shop.ata.org.au/shop/pearscollection
27. Pears, A (2017) *Poor households are locked out of green energy, unless governments help*, in The Conversation, August 6 accessed on https://theconversation.com/poor-households-are-locked-out-of-green-energy-unless-governments-help-81987
28. Stengers, I. (2005) The Cosmopolitical Proposal, In *Making Things Public: Atmospheres of Democracy*, ed. Latour, B and Weibel, P Cambridge, MA: MIT Press, 1002.
29. Houston, D and Vasudevan, P. (2018) in *The Routledge Handbook of Environmental Justice*. Holifield, R., Chakraborty, J. & Walker, G. (eds.). New York: Routledge, pp. 241–251.
30. Ibid. p. 245.

CHAPTER 3

Bringing Missing Actors to the Table

Abstract This chapter focuses on how responses to climate change have tended to focus on some actors and sectors to the exclusion of others. Central to this is the idea of *enabling* innovation. Elements of enablement include leadership, relationship building (e.g. trust, commitment, open communication, collaboration and information exchange), skills development and competence building (e.g. problem identification, developing solutions), decision-making support (e.g. financial or human resourcing, delegating power to act) and response-ability (i.e. to external disruptors, e.g. storm events or financial shocks, risk management, entrepreneurial readiness, experimentation). Activist practices that map onto these five dimensions of enablement illustrate how each dimension offers a different way to bring missing actors to the table. New ways of bringing diverse localised actors together (e.g. trade unions, religious organisations and businesses) can break through barriers to collective action, ushering in new forms of climate activism.

Keywords Enabling • Innovation • Experimentation • Disruptors • Power • Response-ability

W. Steele et al., *Quiet Activism*, https://doi.org/10.1007/978-3-030-78727-1_3

ENABLING SOCIALLY INNOVATIVE PRACTICES

Climate change has been described as the greatest challenge humanity has ever faced, an opportunity, a threat, a wicked problem, a moral issue and even a potential "planet killer". It is little wonder then that some people find the idea of responding to climate change to be beyond their capacity to act, too scary, or even unnecessary.[1] Part of the tremendous difficulty in building collective climate change action is that many people feel that it is either not their responsibility or that they are unable to participate.[2] As expressed by participants in our research interviews some of the barriers identified include the fear that they don't know enough, confusion about where to begin and for some the nagging feeling that any action they take could feel trivial in the face of the global climate emergency crisis. And for some vested interests, collective action is anathema. They have much to gain by ensuring the status quo continues.[3]

On the surface, it is easy to despair about the slow, blundering and often paradoxical institutional and policy responses to climate change. In Australia for example there are three levels of government—Commonwealth (national), state (provincial) and local (municipal). The Commonwealth government's response has been patchy. Gains made under one political administration have been eroded or reversed by another. At the state government level similar trends are evident. Some states have made considerable progress (e.g. the Australian Capital Territory), whereas others have lost ground after early advances (e.g. Tasmania). Although many state government agencies have been progressing a wide array of actions (e.g. education, research, community grants), they have mostly been disjointed or incremental.[4] At the local government scale, there has been substantial action, mainly among better resourced municipalities some of whom have formally declared "a climate emergency", but at times this too has been contested. It is not uncommon for a climate change strategy to be prepared under one mayor only to have it thrown out by another, following a council shakeup after a local government election.

Three problems appear to shape how Australia responds to climate change. First, governments, who are often seen as the most appropriate to lead action, are constrained by the highly prescriptive legislative frameworks in which they operate. They act according to strictly defined roles and responsibilities. For local government, these roles and responsibilities tend to focus on service delivery (e.g. waste management), regulation (e.g. parking enforcement) and community health and safety (e.g. immunisations, food safety certification). Because there is no constitutional recognition of local government in Australia, councils exist 'at the pleasure of the state'. In practice this is a very hierarchical structure—state governments tend to decide what to do, and then tell local governments how to do it. Sometimes the inter-governmental relationships that follow can be highly adversarial.

A second problem is that much of the language used in response to climate change is framed around risk and responsibilities—which makes many actions seem rigid, cold, impersonal and highly technocratic. Long tables, spanning more than ten pages, of identified risks, level of risk, departments with responsibility, asset managers and prioritised budget responses and timeframes for implementation make for hard reading. They can also paralyse action; the tasks ahead can seem immense, expensive and beyond the capacity of a single organisation. This is further complicated by vested interests and industry/party political lobby groups who seek to obfuscate the policy process and stymie the real action needed to meaningfully address the climate emergency.

A third problem, which we focus on here, is that many actors that could potentially play a stronger role in climate change adaptation and mitigation tend to slip through the cracks in governance structures or find themselves on the margins and denied a seat at the table. This is partly due to a certain level of distrust between the key actors. Local government officers can be wary of community activists who lobby their elected representatives, apply political pressure and grab local newspaper headlines. Community members can feel that councils are inordinately slow, bureaucratic and unapproachable. Businesses can feel that they are stymied by red tape. Aboriginal people are seldom considered by local government with respect to climate change response. Indigenous ways of knowing and doing do not readily align with colonial models of governance, and they are readily discounted. This is despite the fact that many Aboriginal communities will be impacted hardest by coming changes.[5] And some elected representatives may worry about moving too fast, lest they lose their mandate on which they were elected.

In considering whether there might be a different approach to climate change response in Australia, we have developed a framework of practice principles for quiet activism at the local scale. Laura Pottinger's work on quiet activism is instructive. Pottinger helps to shift our attention away from the structures, institutions, agencies and power dynamics that have garnered so much scholarly consideration[6] to humbler 'acts of kindness, connection and creativity'.[7] Indeed, Pottinger defines quiet activism as: 'modest, embodied acts that often entail processes of production or creativity, and which can be either implicitly or explicitly political in nature'.[8] Central to her framing is the idea that activism can include 'small acts, such as the creation of interpersonal connections that construct social networks'.[9]

In the previous chapter our emphasis was on building and bridging the knowledge base, and the power of local stories and practices purposefully focused on climate action. Developing this further, key questions we consider in this chapter include: who are the so-called missing actors in climate change responses, what sorts of responses are they taking, and what lessons can we learn from them? In what follows, we examine the roles that these less appreciated actors are playing and how they are "bringing themselves" to the table. As noted earlier in this book, there are some actors who are noticeably absent from our stories and analysis (e.g. Aboriginal people)—not because we excluded them, but because it was not possible to engage with them given our timeframe, skillsets and social connections. We recognise this is an important limitation of our cases and point to the climate justice practices of First Nations organisations such as *SeedMob* whose networks support the local work of Indigenous climate activists.[10]

Climate Action: A Seat at the Table

Central to our thematic focus and analysis in this chapter is the idea of *enabling* innovative practices[11] as a form of quiet activism. Innovation might be conceptualised as occurring across three domains—behaviour change, institutional change and built form change—such as tree planting, lighter coloured roads and roofs, heatwave shelters, stormwater harvesting and so on. Elements of enablement include (i) leadership, (ii) relationship building (e.g. trust, commitment, open communication, collaboration and information exchange), (iii) skills development and competence building (e.g. problem identification, developing solutions), (iv) decision-making support (e.g. financial or human resourcing, delegating power to act) and (v) response-ability (i.e. to external disruptors, e.g. storm events or financial shocks, risk management, entrepreneurial readiness, experimentation).[12]

Activist practices that map onto these five dimensions of enablement illustrate how each dimension offers a different way to bring missing actors to the table. Quiet acts of caring and inclusion can build capacity for action, thus engendering response-ability. New ways of bringing diverse localised actors together can break through barriers to collective action, ushering alternative forms of climate change response. Employing a conventional understanding of enablement, Brooks suggests that enablers of innovative responses to climate change can be assigned to three types—policies, institutions and technologies.[13]

A similar understanding identifies not only access to information, improvement in science, institutional reform, and windows of opportunity as enablers, but also collaboration, building bridges and trust as enablers, offering a tantalising clue that social innovation may play an important role in enabling climate change response.[14] Redolent of this second conception of enablement is enabling from a health-care context; they see enablement as a "process" of *empowerment that derives from capacity building*.[15] In much the same vein, in their assessment of nursing practices of care, Frost et al. argue that enablement relates to developing understanding, acquiring the ability to cope and becoming self-sustaining.[16] Elements of enablement here take on a more affective dimension.

The local-scale stories and practices that map onto these five dimensions of enablement illustrate how each dimension offers a different way to bring missing actors to the table. For instance, developing political agreement about the broad principles of a local government strategy (e.g. by putting the technical details in supporting documents) can bring politicians to the table to build effective leadership without engendering partisanship or getting stuck in technicalities. Fostering community development practices (e.g. via community multicultural days to promote social inclusion) can engender capacity for adapting to climate impacts, and increasing resident involvement (e.g. neighbours checking on each other during extreme weather events), can help build response-ability. Finally new ways of bringing diverse actors together (e.g. artists, scientists, bankers and gardeners) can build stewardship or solidarity which can help to break through some existing barriers to action. We now turn to several examples from our research in the Australian context: the former NSW Nature Conservation Trust, Climate Risk's Karl Mallon and the non-profit organisation based in the Gold Coast "Gecko", to illustrate the diverse range of quiet activist practices in this theme.

NATURE CONSERVATION TRUST: CARE AND PROTECTION

Non-human species (plants and animals) seldom get a "seat at the table" when it comes to climate change responses. Where plants are considered for example, it is often in the context of their instrumental value (e.g. how "green infrastructure" can cool cities or mangroves can limit coastal erosion) or in the language of risk reduction, because vegetation is seen to pose a risk to people and property from storm events or bushfire. Animals too are usually ignored in climate change responses, unless they are (i) *victims* (e.g. the bleaching of the Great Barrier Reef or the death of thousands of flying foxes during heatwaves), (ii) *vectors*—sources of disease transmission (e.g. mosquitos and Ross River virus) or (iii) *bellwethers*—because, for example, their presence in new areas signifies a startling climate shift (e.g. tropical Queensland fish finding their way to the coastal waters of Tasmania). Often the images of plants and animals in the context of climate change evoke loss and mourning, such as drowned cattle following a cyclone, orphaned koalas following ravaging drought and scorched ancient alpine vegetation following uncontrolled wildfire. Such evocations can lead us to feel helpless and paralysed. But it doesn't have to be this way.

As far back as the 1970s, scholar-activists began posing questions about how we can better include plants and animals in decision-making, by giving them a voice and/or legal standing.[17] More recently, there have been a number of prominent international decisions granting personhood to non-human entities such as rivers.[18] There is already a precedent for this in society, where corporations oftentimes enjoy the same sorts of legal protections as people. Increasingly, there is a recognition that although climate change will affect plants and animals in profound ways, action to help them survive and even thrive is still possible—and is critically important. The actions of the former Nature Conservation Trust, for instance, show what can be achieved through empowering local communities to act to protect, and care for, their land, in a model of stewardship for present and future generations.

The Trust was created as an independent organisation by the NSW Parliament in 2001 with the passage of the *Nature Conservation Trust Act*. It was responsible for conserving and enhancing natural heritage on private land in NSW and educating the public on the importance of the varied places, plants and animals that make up this rich natural heritage. The Trust helped build coalitions of farmers and landowners, enabling them to take small-scale actions that collectively scaled-up into bigger results. The story of the Nature Conservation Trust (hereafter NCT) is significant because it highlights how organisations working as brokers between government and community can bridge the sorts of divides mentioned earlier, and thus enable citizens to protect unique and valuable natural heritage. Although in 2016–2017 the Nature Conservation Trust was replaced by the Biodiversity Conservation Trust (BCT), the BCT will continue to administer agreements concluded under previous legislation with the Nature Conservation Trust.

Since its creation, the NCT worked with private landholders and communities to protect over 55,000 hectares of land with in-perpetuity covenants across New South Wales. These in-perpetuity covenants are legal agreements that permanently protect significant vegetation and treasured places from development. In-perpetuity covenants, once applied, protect areas of natural heritage by restricting development, vegetation removal and interference. A variety of different conditions are available in these covenants, allowing for flexibility in land-use and dynamic land management.

Unlike nature reserves, the land remains in private ownership and negotiated agreements allow its use for different purposes. The NCT has worked with local landholders, organisations and government agencies to identify areas of important natural heritage. A range of factors are considered including: biodiversity, general species richness, vegetation communities which are poorly represented within existing conservation reserves, threatened species habitats and ecosystems that contribute to greater climate change resilience and adaptation. By protecting larger areas of vegetated land, non-human species have a better chance of adapting "in-place" to changing climates and climate variability, or of moving through vegetated corridors to other places that may better suit their needs. An example is the Capertree Valley.

The Story of Capertree Valley

The Capertree Valley, located 135 kilometres northwest of Sydney, is internationally renowned for its birdlife and biodiversity and is home to an array of World Heritage Sites. A local landowner, we'll call her Julie, had spent years planting significant numbers of trees on her property in the Capertree Valley, creating new habitats and enhancing local biodiversity—acts of quiet activism. Concerned by the lack of legislative protection for her trees if she sold the property, Julie approached the NCT to see what they could do to help.

The NCT conservation covenants offered Julie the opportunity she sought to protect the place she valued, without impacting its ability to be sold or used for sympathetic purposes in the future. Working together with the NCT, Julie was able to apply in-perpetuity covenants to her land. Once an agreement like this was developed, the NCT provided regular contact and ongoing support in the form of management guides, property visits and suggestions for additional actions. Together with identifying effective land management practices and developing a management plan, the NCT provided landholders like Julie information about how their properties fit within the overall conservation "mosaic" of New South Wales. They could then see how applying covenants to their property helped to create and secure state-wide assemblages of protected habitats.

Giving important insights into her quiet activism, during the process of protecting her land, Julie leveraged her local networks, connecting fellow community members with the NCT and built knowledge and awareness about the possibilities of using covenants to protect larger areas of natural heritage in the valley. The subsequent uptake in interest by Capertree locals and the engagement of organisations like the local Catchment Management Authority led to better conservation outcomes in the Capertree Valley. The NCT effectively empowered a local coalition of actors to create an ongoing project of care and response-ability, forming part of a larger network of protected areas in New South Wales.

Julie's story highlights how the NCT, through processes of engaging individuals, communities and organisations and empowering them to act to protect locally important plants and animals, didn't just provide a pathway for locals to protect their landscapes; the organisation also offered ongoing support in conserving and maintaining bushland, forests and other significant habitats. In this way the NCT employed several dimensions of "enabling" to foster a climate change-adaptive response. The NCT built upon the existing ethics of care in the community to foster ongoing relationships and network building. The NCT scaled up actions, from local landowners, through regional groups, to a state-wide network. Using its effective practices of conservation education, awareness and action, the NCT enabled individuals and groups to contribute to larger scale regional conservation initiatives and to improve awareness and engagement in other communities.

When organisations and governments provide clear pathways for communities to act, innovative and impactful actions can be achieved. Bringing a diverse and motivated public to the table and enabling them to scale up local actions can build networks of action and foster champions—building the resilience, response-ability and human and non-human relationships required to better respond to changing climates. Another example is the work of Karl Mallon and Climate Risk Australia.

Addressing Risk and Disruption

One of the major challenges with missing actors is how to bring climate itself to the table. Although seldom explicitly recognised, climate has agency—the capacity to act.[19] While we cannot sit down in a conversation with "climate", climate change manifests in all sorts of ways on a daily basis—the stifling temperature in a heatwave, smoke choking us from a nearby bushfire or the flying foxes lying dead under the trees outside our houses. It is, however, possible to employ innovative practices to 'make climate change manifest' in other ways. In this second story, we consider the work of Karl Mallon and how he is seeking to help homeowners better engage with climate change and climate variability, and better understand the impacts of climate change on where they live—thus bringing them to the table. In a sense, Karl acts in the role of "broker", helping to bring together actors who may previously have not sat together at the same table, in climate response. Karl sees his work as part of a process of disruption.

Disruption can have both negative and positive impacts—with the potential for renewal and adaptation existing alongside threats of breakdown or stagnation. Climate change is already disrupting economies, governments and communities in a multitude of ways and will continue to do so on an increasingly larger scale. Karl Mallon's particular brand of disruption, which emphasises collaboration and the public interest, is fostering changes in how the broader public and the financial sector discuss climate change—effectively bringing missing actors to the table by bridging information gaps for homeowners and home purchasers. By bringing banks, insurers, homeowners and purchasers into a discussion of climate change's tangible impacts, Mallon argues that greater awareness and action can occur, leading to innovation in industries and shifts in not only values but also in property *valuations*.

Karl Mallon has been working in the climate change policy and technical space for more than 25 years. Mallon is Director of Science and Systems at Climate Risk, a consultancy specialising in climate risk adaptation software and specifically dedicated to climate change adaptation in Australia. He is also Director of Science and Engineering at XDI, a consultancy providing integrated climate risk and modelling reports. Karl specialises in identifying climate change risks to infrastructure, providing services to local government, as well as insurance and other financial industries. Mallon's work integrates knowledge of science, policy, finance and IT sectors to develop tools for the general public, government and private sector, to inform and improve decision-making.

For example, Mallon's Climate Valuation website supports and improves the public's decision-making capacity by providing accessible reports that integrate a variety of data on sea level, storm and inundation risks. In addition to his consulting work, which he sees as building the public's decision-making capacity on climate change, Karl Mallon has published a range of editorials and articles in the *Sydney Morning Herald* and other media outlets, has participated in panels and interviews, as well as provided modelling and submissions to Commonwealth Bank and Productivity Commission reports.

As a self-described "disruptor", Mallon has advocated for the public release of information, data and models that are presently only available to large insurers and organisations. He has identified a variety of benefits that can be gained by decreasing the information asymmetry between organisations and the public. In his work, Mallon argues that a public release of this information would improve the social licence to operate of banks and insurers, currently facing substantial public distrust, thus reducing risk to their business models and building awareness and desire for action among the general public. His work engages in the monetary language that underpins much of the public debate in Australia about climate change response, enabling a type of translation where climate change impacts and risks can be "made real" for homeowners.

Climate Valuation: The Website

In 2015, the Productivity Commission released a report on natural disaster funding in Australia. In this report, the Commission identified significant information asymmetry or knowledge gaps between the public and industry, as well as low participation of insurers in public and household discussions around risk assessment and the impacts of climate change. The Commission found that both insurers and governments should take steps to communicate information around hazards to households along with other initiatives.

Climate Valuation works to disrupt this current knowledge gap and industry status-quo through the provision of risk and hazard information in an accessible format 'that an ordinary person can use when they make a decision'.[20] According to Mallon, the Climate Valuation website uses climate and environmental modelling and risk data to provide 'an important input to the total information puzzle for home-owners, potential purchasers, lenders and insurers'.[21] Using the website, anyone can enter an address to generate an individual property report, describing the risks of flooding, sea level rise and an assessment of vulnerability and potential adaptation options. The website makes available a number of reporting tools to determine the risk of coastal inundation, its value impact and an assessment of the "why" of vulnerability as well as potential adaptation options.

When asked why he founded the organisation hosting this service, Mallon said: 'Essentially, the landscape was unfair—people were acting in ignorance ... they weren't necessarily being exploited, but they certainly weren't being looked after'.[22] For Mallon, Climate Valuation combats "information asymmetry" between large organisations and homeowners. By building public awareness of the financial risks of climate change, it can incentivise everyday homeowners and property purchasers to act on reducing the risk they face, such as borrowing money to finance home renovations that can lessen fire hazard, improve thermal efficiency or reduce flood exposure.

Mallon describes the website as a way to enable citizens to further build their activism around climate change. He says: 'we can start to put information into people's hands so they can make informed decisions. There is nothing greater than a wallet in driving behaviour change'.[23] Through providing property reports that identify risk in a highly accessible format, Mallon enables homeowners to make better decisions and to begin to minimise risk exposure and vulnerability.

In an interview on Australia's Radio National, Mallon highlighted how the Climate Valuation website is being used in conversations with banks to encourage them to provide mortgage options for household adaptation measures—thus increasing revenue, ensuring long-term "insure-ability" and protecting the interests of homeowners and purchasers.[24] In turn,

Mallon notes that this may lead homeowners to rethink valuation, risk and purchasing decisions as well as to agitate for action on climate change with financial service providers. The recent Commonwealth Bank 2018 Annual Report, featuring Climate Risk work, presents an example of this greater discussion of the risks of climate change, and shows how the work of Mallon is beginning to leverage greater involvement from the private sector in climate change response.

Missing Actors, a Shifting Designation?

In a 2018 interview, Mallon discussed how a number of insurers and banks have come to the table and are now more cooperative in approaching the challenges of climate change. He argued that banks and insurers are now dedicating serious resources and time to seeking to better manage risks of climate change, for example reducing fiduciary liability, and are now collaborating with a greater range of actors. An important question remains though: who are the missing actors to bring to the table in discussions around risk, value and climate change?

Mallon underscores the need to build awareness of the impacts of climate change in groups beyond highly informed audiences. In particular, he identifies residents of especially vulnerable areas, such as coastal and low-lying regions, as needing better education about the risks they face. Bringing these residents to the table, he suggests, would encourage better financial decision-making and further awareness and engagement with climate change.

Karl Mallon's work highlights the important role of disruption 'with the door still open'. By combining disruption with an openness to collaboration, Mallon is bringing some key missing actors to the table (e.g. banks) and in so doing is promoting a better awareness of, and engagement with, climate change. Karl's work also demonstrates the necessity to reflect on which actors are missing and how to bring them to the table—something that changes over time and in different contexts. In his work with Climate Valuation and Climate Risk, Mallon has also built bridges between expert knowledge and homeowners, advocating for greater openness, dialogue and collaboration between insurers, banks and their clients. Mallon's work has helped to support and to empower consumers in decision-making, as well as to bring insurers and banks to the table in highlighting the significant benefits of open data access and collaboration around asset protection.

This work doesn't necessarily directly benefit marginalised and disadvantaged residents (e.g. tenants, poorer people, homeless), and presently, it is not reducing the cost of insurance.[25] But it has the potential to reduce indirect financial impacts that occur when the state has to rescue people from places impacted by natural hazards or bail out property owners after a disaster. Mallon's work thus lessens some of the financial burden of climate change on broader society—through skills development and competence building that form key elements of enablement. The next story takes these types of initiatives one step further. Where Mallon's work is focused on the present generation, and humans, members of the non-profit organisation called GECKO, based on the Gold Coast, are actively caring for non-humans and future generations through multiple acts of quiet, and not so quiet, activism.

Bringing Local Community to the Table

Residents, local government officers, state government officials and community activists often express the view that not enough is being done by the broader community to respond to climate change. Many climate change adaptation plans, for example, point to the need for better community engagement and more participation, but few identify how this can be achieved. Recognising that within towns and cities there are multiple communities, with different identities, capabilities and capacities, and finding ways to bring community to the table can be a challenge.

Where "the community" is recognised as a legitimate actor in climate change responses, it is often seen as a repository of local knowledge. Local people are believed to have been living with climate change already, so they are seen to be naturally adapting. For instance, the International Institute for Environment and Development points to a wealth of knowledge about adaptive responses purportedly held by the community, who can use their existing decision-making processes to adapt.[26] But exactly what these processes are, how they can be effective and how diverse groups of people can identify and agree upon adaptation pathways for action are seldom specified.

Building and strengthening relationships within and between different segments of society can enable innovative responses to climate change and climate variability. The work of the *Gecko Environment Council Association*—formerly the Gold Coast and Hinterland Environment

Council—is an example. Gecko is a community conservation organisation operating in the Gold Coast (Queensland) region. It was formed in 1989 following the merger of six local conservation groups. Acting initially as a clearinghouse of environmental knowledge and coordinating diverse local environmental initiatives, Gecko evolved to become a focal point for grass-roots activism and actions to protect and regenerate natural heritage as well as to promote more ecologically sustainable practices on the Gold Coast.

Gecko works across a variety of campaigns and local actions. These include knowledge brokering, grant writing, community organising, campaigning around local and national environmental issues (e.g. Stop Adani), representing the community and environment's interest via membership on state and local government committees, and engaging in activities such as revegetation, environmental education and supportive practices like backyard food growing. Gecko also has two social enterprises—*Regen Australia*, which is an environmental consultancy, native plant nursery and ecological restoration enterprise, and *Gecko Ed*, which offers environmental education services.

Gecko's diverse engagements with the community have led to a variety of initiatives that could be termed quiet activism, such as the 2016 *Climate Change for Good* conference and community art projects. Through these diverse actions, Gecko has been successful in raising community awareness about climate change impacts. Its *Eat your Backyard* campaign, providing information and effective practices about how to grow your own food at home, is a good example. In the words of one of its members:

> we certainly don't take rallies and things like lightly, that's because it is enormously energy draining, and takes you away from lots of other things that you want to be doing as well. So campaigning and rallying at that level are last resorts. …'Eat your Backyard' was a successful event, engaging people in growing food in small spaces. We had about 300–400 people come through … in addition to the speakers we had stalls, like the regional bee-keepers and OZ Harvest, butterfly people, a few others … it can be developed further.[27]

Reaching beyond homeowners to renters, children and sometimes marginalised members of society, Gecko is working to make climate change manifest. Instead of being a scary future event, Gecko has begun the work of helping residents to better imagine what sea-level rise looks like now, and in so doing, to prepare for this eventuality. The *Rising Blue*

Line project is another example of how Gecko is bringing missing actors to the table through quiet activism.

Rising Blue Line

The Rising Blue Line was a 2017 installation on the banks of the Currumbin Creek by the artist David Paynter and fellow Gecko volunteers. Trees, poles and posts along the creek in Winders Park were wrapped with blue recycled towel rolls to highlight the projected high-tide mark in the year 2100. Paynter worked alongside over 500 Gecko volunteers during the SWELL Sculpture Festival, September 2017, to install the project. Paynter re-imagines climate change as 'providing us with the opportunity to do things differently',[28] and asserts that through connection with each other, nature and future generations, residents can respond to the need for urgent changes.

The Rising Blue Line installation combined citizen science, public art and climate change knowledge in an innovative way (see Fig. 3.1). In basing its work around collaboration and community, Gecko embodies what Carl Boggs describes as *prefigurative politics*: 'to prefigure is to anticipate or enact some feature of an "alternative world" in the present, as though it has already been achieved'.[29] Making manifest the physical impact that rising sea levels will have on the Currumbin Creek landscape, the installation also prefigured the relationships and collaborations that will be required if locals are to adapt in place. The open, collaborative and community-focused actions being led by Gecko reflect the organisations' embedded philosophy of community activism that makes a difference. But Gecko has also been working in other ways, which are effective in bringing them to the table.

Building Local Community Activism

Increasingly locked out of local government by a conservative political elite, in many ways, the people who comprise Gecko have 'made their own table and sat themselves down at it'. One member of Gecko, who we interviewed, told us:

> On a local level our President has been going around meeting all the councillors, particularly the new ones, just to let them know what Gecko is about and for them to see a... different face to just the campaigning. There are

Fig. 3.1 Rising Blue Line at SWELL Sculpture Festival 2017. (Source: Gecko Environment Council, https://gecko.org.au/getinvolved/gcan/slr/2017/10/ rising-blue-line/)

some in Council that would not talk to us because they think we are campaigners. So that is a kind of different way in, emphasis[ing] the education and the events side other things, which they [councillors] find easier to accept, without realising that we actually use the events to do our campaigns![30]

Members and volunteers of Gecko are not just fellow activists but more importantly are fellows and friends—practising radically situated acts of caring. For example, writing about her work, Cate Ware, a Gecko member, has stated: '(f)riendship, community, skills development, personal satisfaction of supporting something that makes a difference and a lot of LAUGHS! What I do changes me and the world!'.[31] For Cate and other members of Gecko, this community-based approach centred upon personal relationships is a key ingredient to their effectiveness. The organisation prides itself on providing a space to listen, discuss and learn about climate change.

Monthly social meetings with guest speakers including scientists, academics, artists and activists bring locals and non-locals together to discuss wide-ranging topics including non-human species welfare, human-nature relations and climate change response—exemplifying quiet activism. Speaking about their *Climate Change for Good Conference*,[32] one of our interviewees said:

> Gecko members have been aware of the issue of climate change for eleven or twelve years, and we've basically informed ourselves on most of that period of time by going to conferences, and discussion groups, and seeing films and all that kind of thing ... So about two years ago we decided we'd had enough, and that our government wasn't going to do anything useful and we had to do something ourselves within that sphere of influence in the hope that it will fit into the national and global actions that are happening. And we felt that if we did it at our kind of level, then hopefully that will also influence our decision-makers.[33]

In building a community around activism and solidarity, Gecko volunteers and members have forged strong links with one another and the broader community. Acting through the relationships they build, Gecko members motivate and inspire—drawing upon opportunities to mobilise around locally important environmental issues (e.g. tree clearing, wetland draining) to build a broader social and environmental constituency, with the goal of achieving 'a vibrant Gold Coast Community where people and animals, plants, water, air and earth all form a healthy, harmonious system'.[34] Much of the work of Gecko mirrors the different dimensions of enabling. Through its innovative local practices, the organisation imagines and realises an alternative climate change adaptation pathway for the local Gold Coast community—one based on collaboration, exchange, trust, commitment and most importantly, caring.

Gecko is a non-profit organisation that is deeply embedded within local places and communities. Focused on prefigurative actions, it builds enduring relationships, based on trust, openness, honesty and a commitment to action, that drive its climate change adaptation work. Values of friendship and solidarity weave together all areas of Gecko's work, from political campaigning to the quiet activism of its creative and non-threatening practices (e.g. eat your own backyard). Integrating relationships within the work of the organisation allows it to foster compassion, engender reflective discussion and achieve community buy-in, thus bringing artists, students, activists, retirees, people in between jobs and other missing actors to the table. These are examples of the quiet but deeply enabling actions we can take when the foundations of how we live, what we do, and what we feel are irrevocably destabilised in the contemporary climate of crisis and change.

CLIMATE RESPONSE-ABILITY

In this chapter we have considered how community groups, entrepreneurs, artists, ordinary residents and plants and animals are interconnected through acts of quiet activism to advance responses to climate change. Importantly, the framings of climate change that these diverse groups of actors employ are not the typical risk, threat or even opportunity framings that are common in media, government and scholarly discussions of climate change. Rather, these actors have used different framings to enable action—based upon loss, love, nurturing, compassion and solidarity. They appear to engender responses that foster a sense of possibility, if not hope for the future. Their understandings are sophisticated, deeply personal and personally political—and they show that they are acting with intent in ways that span scales from the body to the planet. Many have shown a willingness to go around, underneath, over or through government institutions—and sometimes even working with institutions—to effect change.

The connective thread, weaving these various stories together, has been the idea of enabling social innovation—via conditions, thoughts and actions. The innovations that these actors have explored and implemented span behavioural, institutional and physical domains. Gecko has used art as a way to make possible future sea levels visible to residents in a tangible, but non-threatening way. Karl Mallon has developed a tool to help prospective homebuyers to be aware of the various ways that climate change—as an actor—is, and will be, an ever-present feature of where they choose to live. The Nature Conservation Trust has provided a way for non-humans to potentially adapt-in place, through the commitment and stewardship of landowners. One of the ways that these missing actors are making a difference is how they bring alternative ways of knowing and doing into their climate change responses. They may lack the power of vested interest groups, may lack the organisational strength of trade unions, may lack the financial clout of the insurance and banking sectors and may lack the numbers of organised religion, but they enact other forms of agency.

This is not to say that these strategies and actions employed by these actors are perfect—they are not and could not be—it would be naïve to think otherwise. The individuals and groups we have considered here certainly do not see themselves that way. These so-called missing actors and their various strategies are not about "saving the world", but they do show us that making a difference is possible. There are some important take-home messages from these stories that emphasise the power of quiet activism to bring missing actors to the table. Perhaps most importantly, these participants did not wait to be invited to work on climate change responses, rather they brought themselves to the activist table. In the following chapter we seek to further this focus on enabling and regenerative practices by turning to the third theme of our quiet activism framework—walking together with care.

NOTES

1. Lucas, C.H. and Davison, A. (2019) Not 'getting on the bandwagon': When climate change is a matter of unconcern. *Environment and Planning E: Nature and Space*, 2(1), pp. 129–149.
2. van Valkengoed, A.M. and Steg, L. (2019) Meta-analyses of factors motivating climate change adaptation behaviour. *Nature Climate Change*, 9(2), pp. 158–163.
3. See Hornsey, M. and Fielding, K. (2020) Understanding (and reducing) inaction on climate change. *Social Issues and Policy Review*, 14(1), pp. 3–35.
4. Byrne, J., Gleeson, B., Howes, M. and Steele, W. (2009) 'Climate change & Australian urban resilience: The limits of ecological modernisation as an adaptive strategy' in S. Davoudi, Crawford and Mehmood (eds) *Planning for Climate Change: Strategies for Mitigation and Adaptation*, Earthscan, London.
5. Nursey-Bray, M., Palmer, R., Smith, T.F. and Rist, P. (2019). Old ways for new days: Australian Indigenous peoples and climate change. *Local Environment*, 24(5), pp. 473–486.
6. See for example Tompkins, E.L. and Adger, W.N. (2005) Defining response capacity to enhance climate change policy, *Environmental Science and Policy*, 8, 562–671; Göpfert, C., Wamsler, C., and Lang, W. (2019), Institutionalizing climate change mitigation and adaptation through city advisory committees: Lessons learned and policy futures, *City and Environment Interactions*, 1, 1–12.

7. Pottinger, L. (2017) Planting the seeds of a quiet activism, *Area*, 49(2), 215–222.
8. Ibid., p. 216.
9. Ibid.
10. See https://www.seedmob.org.au/
11. See for example Burch, S. (2010) Transforming barriers into enablers of action on climate change: Insights from three municipal case studies in British Columbia, Canada, *Global Environmental Change*, 20, 287–297; Neumeier, S. (2017), Social innovation in rural development: identifying the key factors of success, *The Geographical Journal*, 183(1), 33–46; Scott-Cato, M. and Hillier, J. (2010) How could we study climate-related social innovation? Applying Deleuzian philosophy to Transition Towns, *Environmental Politics*, 19(6), pp. 869–887.
12. Pasquiani, L., Ziervogel, G., Cowling, R.M. and Shearing, C. (2015) What enables local governments to mainstream climate change adaptation? Lessons learned from two municipal case studies in the Western Cape, South Africa, *Climate and Development*, 7(1), 60–70.
13. Brooks S. (2014) Enabling adaptation? Lessons from the new 'Green Revolution' in Malawi and Kenya, *Climatic Change*, 122(1–2), 15–26.
14. Jones, J., Winch, S., Strube, P., Mitchell, M., Henderson, A. (2016) Delivering compassionate care in intensive care units: nurses' perceptions of enablers and barriers, in *Journal of Advanced Nursing*, 72(12), pp. 3137–3146.
15. Hudon, C., Tribble, D., Bravo, G., and Poitras, M-E. (2011) Enablement in health care context: a concept analysis, *Journal of Evaluation in Clinical Practice*, 17, 143–149.
16. Frost, J.S., Currie, M.J., Northam, H.L., Cruickshank, M. (2017) The experience of enablement within nurse practitioner care: a conceptual framework, *The Journal for Nurse Practitioners*, 13(5), 360–367.
17. See for example Stone, C.D. (1972) Should trees have standing? Toward legal rights for natural objects. *Southern California Law Review*, 45, pp. 450–501.
18. O'Donnell, E. L., and J. Talbot-Jones. (2018) Creating legal rights for rivers: lessons from Australia, New Zealand, and India. *Ecology and Society*, 23(1) art. 7.
19. Matthews, T., Lo, A.Y. and Byrne, J.A., 2015. Reconceptualizing green infrastructure for climate change adaptation: Barriers to adoption and drivers for uptake by spatial planners. *Landscape and Urban Planning*, 138, pp. 155–163.
20. Quotation taken from Australian Research Council (ARC) DP150100299 interview transcripts
21. Ibid.

22. Ibid.
23. Ibid.
24. see 'Climate Risk' Karl Mallon on ABC Radio National Saturday extra with Geraldine Dooguehttps://www.abc.net.au/radionational/programs/saturdayextra/karl-mallon/9235584
25. Booth, K. and Kendal, D., (2020). Underinsurance as adaptation: Household agency in places of marketisation and financialization. *Environment and Planning A: Economy and Space*, 52(4), pp. 728–746.
26. See https://www.iied.org/climate-change
27. Highlighted in the interviews undertaken in the Australian Research Council (ARC) DP150100299 interviews
28. See Rising Blue Line, accessed on https://gecko.org.au/getinvolved/gcan/slr/2017/10/rising-blue-line/
29. Boggs, C., (1977) Revolutionary process, political strategy, and the dilemma of power. *Theory and Society*, 4(3), pp. 359–393.
30. Ibid.
31. see https://gecko.org.au/about-gecko/our-people/
32. Climate Change for Good Conference, accessed on https://climatechangeforgood.com.au/
33. Quotation taken from Australian Research Council (ARC) DP150100299 interview transcripts.
34. See GECKO website, accessed on https://gecko.org.au/

CHAPTER 4

Walking Together with Care

Abstract This chapter explores the diverse collaborative practices of local climate action. Working together differently draws on a range of diverse knowledge and methods capable of producing generative and creative practices that can create genuinely new opportunities for social and environmental change. In the cases highlighted in this chapter, rethinking action on climate change is identified as a key social experiment and innovation: the ways it can be done; the ways it is being done; and how this generative climate-related activity engages diverse and multiple collaborators at the local scale through other modes, narratives and activist practices.

Keywords Care • Collaboration • Creativity • Narratives • Activist practices • Local action

CARE IN WEATHER-WORLDS

This chapter focuses on the third theme we have identified as part of our quiet activist framework. Walking together with care highlights how socially innovative practices at the local scale can counteract the tensions between planning for certainty and planning for contingency, by turning these around through shared, purposeful community activity. Technical approaches to climate assessment are effective at identifying biophysical

W. Steele et al., *Quiet Activism*,
https://doi.org/10.1007/978-3-030-78727-1_4

risks but they also work to establish normalised patterns for understanding those risks that are associated with fixed ideas about climate, risk, stability and certainty. However, such registers of climate risk do not always recognise the broader cultural implications of the entanglements between humans, weather and climate—what Mike Hulme refers to as the "weather-worlds" that are unevenly made and inhabited.[1]

Against the backdrop of an increasingly unstable climate, diverse weather-worlds are the felt and lived contexts through which people experience climate change. This is about amplifying and mobilising the local maps for negotiating the climate emergency and the lived complexities of inhabiting and walking along together in diverse and changing weather-worlds. This idea of mobilising felt and lived contexts of climate and care is inspired by what we heard from the participants in our research, and also by human geographers Lesley Head and Chris Gibson's call to move beyond limiting frames that narrowly seek to identify climate issues as purely physical phenomena or as an object of environmental and institutional risk.[2]

In the previous chapters we have highlighted some of the stories and practices of quiet activism that are taking place in local Australian communities by enabling localised practices of socially innovative climate action. A key element of this work involves creating spaces of care for people to come together to talk about their experiences, their worries and their hopes. In the bridge-building practices of *Climate for Change,* for example, a Melbourne based not-for-profit organisation introduced in Chap. 2, the aim is to facilitate people's experiences of climate change.[3]

> But I think it's important to provide a service for people to like, for everyday people to have a chance to actually just sit with that, rather than being told that it's a crisis or, you know, the end of the world, or that it doesn't matter, and having to work out somewhere in their busy lives, some position on that, having the time to really sit down and go – "I don't understand this part, could you explain it to me?" or "why are people saying this" or "I'm really worried about that", and sharing that with your friends and family I think is pretty powerful.[4]

This also speaks to the need to bring missing actors to the table. As outlined in Chap. 3 enabling conditions can help us to break through some of the barriers to taking effective action to address the climate emergency. These conditions include: a willingness to be flexible, to experiment, to take chances and to use collective forms of action. These enabling conditions are also premised upon collegial and caring relationships, motivated by affect and concern. Our research participants did not overly dwell on risks, threats and assets—instead they spoke about hope, joy, loss and care. They might be seen as idealistic in some respects, but their actions were deeply practical and were driven by commitment, passion, enthusiasm and even humour in order to address the climate emergency at the local community scale.

Quiet activism draws attention to such entanglements of the personal and the political. Looking into these entanglements sheds light on the role of intimacies, feelings and affects, as well as the unassuming and ordinary acts that contribute to different ways of "walking together" to adapt to climate change. This entails drawing on a range of diverse knowledge and methods and practices capable of producing generative and creative practices that can create genuinely new opportunities for social and environmental change.[5] This in turn foregrounds the diverse ways in which communities of people and practices *become knowledgeable* about the changing and variable urban weather-worlds that we live in. Tim Ingold beautifully illustrates this approach:

> By becoming knowledgeable I mean that knowledge is grown along the myriad paths we take as we make our ways through the world in the course of everyday activities, rather than assembled from information obtained from numerous fixed locations. Thus it is by walking along from place to place, and not by building up from local particulars, that we come to know what we do.[6]

Mobilising, ordinary everyday activities as a means of walking together with care, is critical to our framing of care in local climate activism. Beyond the gentle, generative, acts of care that are central to quiet activism, the practical and ethical commitments and "labours" of care in cities and communities are increasingly articulated by feminist scholars as being a key condition of repairing and responding to climate-changing worlds.[7] It is important to note here that feminist articulations of care do not render it as a means or an end to itself. Care is not a one-directional act transferred from givers to receivers, nor is it a moral imperative to "care" in the same way in all situations.[8]

Rather, care is embedded in (and indeed a conditional of) ethico-political relations.[9] Thus, care sits in productive and creative tension with the responsibilities and practices of *relational wellbeing*. As Maria Puig de la Bellacasa writes, 'a politics of care engages much more than a moral stance; it involves affective, ethical, and hands-on agencies of practical and material consequence'.[10] Building upon this understanding, politics and acts of care in weather-worlds speak to the ordinary intimacies and every-day practices of local climate action. One in which the agencies of care in already existing interdependent worlds are necessarily more-than-human.[11] These small and quiet activisms are interconnected with the more-than-human realities of climate change in diverse ways.

In what follows, we are particularly concerned with how new opportunities and values have been identified and care-fully co-developed as climate action at the local scale. Each of the stories highlights different forms of everyday collaboration between councils, communities, scientists and universities, from engaging local communities in street tree planting to reduce urban heat islands, crowd-funding solar panels and sustainable businesses, to growing community connections through gardening and supporting local foods and the creative arts. The element that ties these projects together is the focus on everyday acts of walking together with care—in homes, gardens, local streets and community centres—and how this can in turn inspire a whole range of new ways of embracing local climate action.

Turning Down the Heat

Climate change and rapid urbanisation are causing air temperature increases across Greater Sydney. In cities and regions, climate change is directly and acutely experienced and understood through adverse weather events such as flooding, wildfires, storm damage and extreme heat. In their report on the catastrophic "Black Summer" of 2019–2020, the Climate Council of Australia reported that 80% of Australians were directly or indirectly affected by the bushfires. Direct local impacts included the following: more than 11 million hectares of land burnt, 33 human lives lost, thousands of homes were destroyed and 3 billion animals thought to have perished during or in the aftermath of the fires.[12] Thousands of holidaymakers were stranded on beaches, highways and in community shelters, while the population of two of Australia's largest metropolitan regions experienced months of breathing in toxic smoke.

In the smoke-shrouded CBDs of Melbourne and Sydney, several large protest actions called on the Federal government to connect the unprecedented fire season with global heating and to substantively act on climate change. In Canberra, Melinda Plesman dumped charred remains from her home in Nymboida, south of Grafton in New South Wales on the lawn of Parliament House with a sign that said: 'Morrison, your climate crisis destroyed my home'.[13] Melinda Plesman's protest was followed by other bushfire survivors who left a 'trail of destruction' using the debris from their burnt homes to mark a trail from Parliament House to the offices of a mining lobby group. While the community protests over the Black Summer, both large and small, were not quiet in either tone or in sentiment, they help to deepen understandings of different types of climate activism/s that we are tracing in this book, and in particular how alongside louder forums of public protest, quieter forms of climate action co-exist and can help to enact new ways of working/thinking/collaborating with care at the local scale.

Blacktown City Council (BCC) for example is one local government area (LGA) that has been working with communities to make them more resilient to climate change. Within the Western Sydney region, local, regional and state governments have been engaging with diverse communities to adapt to urban heat and to "cool the commons". Mitigating and adapting to increasing heat is a key focus of the Western Sydney Regional Organisation of Councils (WSROC) *Turn Down the Heat Strategy and Action Plan*. Western Sydney is a geographically large, economically and ethnically diverse suburban region. Blacktown is expected to experience an additional 5–10 hot days over 35°C by 2030 and an additional 10–20 hot days by 2070 and is working with WSROC on the *Turn Down the Heat Strategy and Action Plan*. The strategy helps to increase public awareness and to facilitate a broader and more coordinated response to urban heat—including how the BBC can address urban heat in its planning approaches and in its Local Environment Plan (LEP).

Since 2018, Blacktown Council has initiated several urban heat-related projects. These projects provide collaborative learning and practical opportunities for communities to work towards achieving liveability and inclusivity at neighbourhood, regional and metropolitan scales.[14] The Council engaged Gallagher Studio to pilot the Cool Streets™ project where local residents participated in selecting tree species to plant on street verges. The project successfully combined community engagement and learning about the benefits of street trees with building local climate resilience and creating liveable public spaces. This programme has since been implemented in two other Western Sydney suburbs.

Cool Streets was complimented with several other programmes such as *How You Can Beat the Summer Heat* and *Cool Your Yard*. The former included the publication of a plant selection guide for residents who are considering planting trees and shrubs at home, while *Cool Your Yard* is a recurring community workshop which provides knowledge about the cooling effect of trees and how to plant trees at home. BBC has also partnered with the University of New South Wales' *Citizen Science Urban Microclimatic* project. In this project, local residents used handheld devices to measure temperatures, humidity and wind speeds in their neighbourhoods. They also learned about how different outdoor materials (depending on colour, material and reflectivity) affect surrounding temperatures and contribute to the urban heat island effect.

These projects inform, and are informed by, *Resilient Sydney* which was the first ever urban resilience strategy for metropolitan Sydney. This strategy set directions to strengthen the ability of citizens, communities and governments to survive, adapt and thrive in the face of global uncertainty and local shocks and stresses. Climate is a main focus within the Strategy, and extreme urban heat is recognised as the 'biggest risk in terms of shocks'.[15] Accordingly, cooling homes and streets is a priority in the BBC for climate resilience. Supporting actions on climate include resilient building, access to renewable energy and measuring metropolitan carbon emissions. Turning down the heat involves collaboration at different scales of government as well as experiments with new ways to engage with local communities, who may or may not, see climate change as a priority.

CLIMARTE: ART AND EMOTION

Taking a very different approach to climate action, art in all its creative forms can provoke us, challenge us and potentially help us to break free from the constraints of prosaic life to ask the important "What If...?" questions. Founded in 2010, Climarte is an alliance of art organisations, practitioners, academics and professionals that seeks to create intellectually freeing and non-threatening spaces to explore the complex and intertwined social, environmental and cultural dimensions of climate change. Guy Abrahams, one of the organisation's co-founders, describes how artists can step into the socio-political arena to reflect upon and discuss the state of society and the natural world. Abrahams underscores the important role of art as not simply a record of events, but also as a catalyst for cultural change. Through its diverse exhibitions, workshops and events,

Climarte brings together ordinary people, experts and art-makers to understand, discuss and 'feel their way together' using art as a medium to explore the implications of a changing climate for species, people and places.[16]

Baby It's Hot Outside was an event run by the Carlton Connect Initiative as part of the 2015 Art+Climate=Change festival.[17] The immersive, speculative event asked attendees to imagine themselves in the Melbourne of 2050. After a three-day heatwave of over 47°C, the attendees were tasked, alongside expert advisors, with developing a plan of action to survive this sweltering future. The audience was asked at the start of the work to "abandon everyday-ism", the pragmatism that blinkers us—limiting what we can see, think and feel. In doing so, the audience co-created a space of dialogue, creating interactions that wouldn't happen in normal life.

A series of "What-Ifs" ensued, ranging from what if trees provided our electricity, to what if art and galleries were re-centred in public life and were further engaged with sustainability? In bringing diverse actors to the table, new ideas emerged, hinting at alternative futures and possibilities. Together, the audience, panel and artists constructed new and novel ideas, imaginings and futures. In departing from the everyday, its hierarchy and considerations, attendees were able to remove their blinkers—to share, interact and exchange diverse knowledge and experiences—allowing new ideas and innovative plans to emerge, hinting at transformative potential.

Foregrounding Emotions

Climarte's work at the intersection of culture and climate change departs from the status quo of climate change communication. So often, discussions of climate change revolve around the modelling of sea-level rise, rising temperatures and the calculus of extreme weather events and can catalyse binary oppositions of belief and denial.[18] By contrast, the works and exhibitions featured in Art+Climate=Change foreground emotion and bring climate change into proximity with our everyday lives and care-full experiences. By focusing on emotion, the organisation seeks to improve how artists and the public can engage in productive and enriching dialogue about climate change, respecting diverse and culturally inflected understandings of climate change as quiet activism.

For example, Anne Noble's 2017 exhibition *No Vertical Song*[19] challenges the established vectors of communicating climate change. The exhibition consisted of a series of black and white photographic portraits of dead bees. With each follicle illuminated, gently jagged wings brushing the air, and bodies expanded in size and in fragility, each image mourning the bees pre-humously rather than posthumously—a warning of what may be lost and the yet-to-be-felt feelings this will cause. Noble's work displays one of climate change's countless future impacts, not through analyses of pesticide impacts but through silent images. She magnifies insects that are part of daily life and asks us to mourn and to feel for those non-human species whose futures, like ours, are uncertain (see Fig. 4.1).

While Noble's work instils a deep and contemplative sadness, itself a form of adapting to loss, other exhibitions and events, like *Baby It's Hot Outside*, engaged with different emotional registers, highlighting new ways of acting, and alternative ways of engaging with climate change. Climarte and the diverse art-makers and organisations it brings together recognise the spectrum of emotional responses to climate change—grief, anger, loss, rage, fear, anxiety and hope. They challenge the status quo of climate change communication by bringing both art viewers and art-makers to the table—legitimising diverse knowledge and ways of

Fig. 4.1 Anne Noble, Dead Bee Portrait #12016, Pigment on Canson Baryta paper

inhabiting the world, and in turn highlighting the potential for new ways of being and acting in the presence of climate change.

By emphasising the legitimacy of people's complex feelings in the face of climate change, Climarte highlights the importance of communicating climate change differently. Ultimately, a rise in average temperatures, increased average temperatures, IPCC scenarios, species extinctions and changes to ocean chemistry can be abstract—but dwelling in climate change and the lived experience of hot, sweaty commutes; dry and choked gardens; empty dams and scorched, silent nature reserves are tangible manifestations of possible futures. Climarte not only offers the possibility of breaking through barriers to action by escaping everyday blinkers, prejudices and blockages, but it brings the future to the table. By enabling people to collectively imagine and share diverse knowledge, feelings and experiences, moving beyond everyday-ism and feelings of despair, art festivals open up possible spaces of potential action by walking together with care.

ONE PLANET: CLIMATE ACTION NOW

The equity dimensions of sustainability are at the heart of two very different programmes: the *One Planet* Council Program which brings together community members and local government to walk together towards care-full sustainable living; and *CANWin—Climate Action Now!* an organisation in Sydney that works to foster community-based initiatives that respond to the impacts of climate change and develop community resilience in the face of peak oil. One Planet is a global initiative that is locally grounded whose aims are 'to create a future where it is easy, attractive and affordable for people to lead happy and healthy lives within a fair share of earth's resources'.[20] The One Planet Councils Program in Australia works with BioRegional, an international NGO responsible for the development of the "One Planet Living Framework".[21] The programme follows ten principles to guide 'practical solutions to climate change'. These principles include a focus on culture and community; equity and local economy; health and happiness; land use and wildlife; local and sustainable food; materials and products; travel and transport; sustainable water; zero carbon energy and zero waste.[22]

The City of Fremantle near Perth in Western Australia, for example, launched its One Planet Fremantle strategy in 2014–2015, to set out how it would become a One Planet Council.[23] The concept behind *One Planet* is that it supports a holistic framework that can be embedded and implemented across all areas of Council business. The City of Fremantle supports staff, local business, industry and residents to strive towards a one-planet lifestyle by 2025. The City has committed to a suite of corporate and community targets and has developed a detailed action plan to guide progress towards meeting these targets. This strategy and action plan are to ensure that events and opportunities are offered to the Fremantle community, so that they can come together and celebrate achievements, share stories, and be rewarded for their participation in building the sustainability of the City of Fremantle.

One notable initiative which demonstrates the theme of walking together is the *One Planet FreoMatch* which aims to increase the value of the City's investment into community initiatives by partnering successful applicants with *Start Some Good* to develop a crowd-funding campaign.[24] This is a collaboration model where local people and communities initiate crowd-funding for projects and when the project has reached 50% of its total target, the City will invest the remaining 50% of the funding goal. In 2015–2016, *FreoMatch* supported eight sustainability projects designed with care to bring action on climate change together with the local community.

The projects ranged from an urban farm that utilises coffee ground waste from Fremantle cafes to grow mushrooms, to a programme delivered in schools to help skill young people in identifying opportunities for positive change in our community. The *Eco-Eats project* funded an eco-conscious catering business that promotes healthy eating, sourced native, seasonal, local, minimal-meat products and sustainable packaging, storage and delivery. Another project funded was *Disco Soup*, which was inspired by initiatives in Germany to minimise food waste by raising public awareness around how much food is lost or wasted because it is imperfect.[25]

With a very different focus and organisational structure to One Planet, CANWin—Climate Action Now! Wingecarribee—is a non-partisan community group based in the Southern Highlands of NSW.[26] A central driver of the group was the recognition of the value of working together with care and to this end CANWin has taken inspiration from Mahatma Gandhi's famous dictum: *we must be the change we wish to see in the world*.

The CANWin community brought together a range of stakeholders by running public events, such as speaker nights, film nights and the *Clean Energy Futures workshop*. Volunteers research and prepare information sheets for members and the public on scientific and technical matters that

Fig. 4.2 Kids pledging to "many hands make renewable energy work" 4

affect the sustainability of life on the Highlands. CANWin initiatives include programmes such as Fruit Rescue, Community Exchange Southern Highlands and Repower Southern Highlands. It collaborates with many other local groups on a range of activities that share a vision for a united, resilient and sustainable community (see Fig. 4.2).

One of CANWin's primary ventures was the Transition Shire Wingecarribee (TSW), part of the world-wide Transition Network.[27] TSW works to bring issues related to climate change and peak oil to the attention of locals. TSW functions as a catalyst to support community groups and activities. TSW initiatives continuously strive for a more locally based, simple and mindful community that moves away from the pervasive fossil fuel-dependent lifestyles.[28] CANWin has recently supported the organisation called *Citizens Own Renewable Energy Network Australia—* CORENA Fund.[29]

The emphasis of CANWin was on community-based initiatives that respond to the impacts of climate change. The CANWin energy working group supported the CORENA Fund to connect community groups to partner in a renewable energy Quick Win Project at community-owned facilities. It is a crowd-funding model that helped to place solar panels on a range of community buildings, for example, a childcare centre. CANWin thus actively assisted communities to go off the grid—a people-powered movement to fast-track Australia's renewable energy revolution—'many hands make renewable energy work in Australia'.[30] In doing so the group is enabling the local community to walk together for action on climate change focused on cultivating care-full local community practices.

CARE-FULL COMMUNITY PRACTICES

Care-full community practices focus on how networks of kinship and alliance can assemble through practices of care. 'However, you find out, we'd love for you to join us on the first Sunday of every month' reads the open invitation on the notice board of the Moss Vale Community Garden (MVCG) website (see Fig. 4.3).[31]

The garden began in 2004 by 'Together in the Highlands' community group with the shared goal of living together sustainably through gardening and planting edible fruits and vegetables. In 2008, the Garden became known as MVCG to help 'people create their own space'.[32] It was hoped that it would be a socially and environmentally innovative community garden space that could be replicated throughout the central Highlands. Specific activities and workshops include grafting, compost making, worm harvesting, cheese making, seed-raising, seed-saving, plant propagation and anything (e.g. basket making) that relates to the garden and to farmers.

Fig. 4.3 'People Create their own Space'. (Source: Moss Vale Community Garden Logo)

MVCG has grown into an innovative platform for both young and old who can socialise through gardening and work together over a home-made pizza. Since 2016, the MVCG has started to move "beyond fences" and to cultivate lands outside its premise with plants and vegetables.[33] MVCG is run by volunteers of all ages and abilities who want to dig their hands in the dirt. The volunteers also grow their own food. Spare produce from the garden is sold to the wider community and garden members also propagate seeds and plants that are local and have heritage value. Seeds and plants are sold in the market every fourth Sunday of the month. MVCG is committed to recycling and reusing everything possible within its activities and have developed an education centre for the local area, and promotes the importance of caring for the environment, recycling and reducing waste and, with a permaculture influence, demonstrating how to grow and enjoy food together at the garden and at home.

An ethic of care-full[34] community practice such as those outlined above illuminates some of the many ways in which everyday practices of care are essential to 'maintain, continue, and repair our "world" so that we can live in it as well as possible'.[35] As an ethical activity and practice, care is per-formed publicly and privately across space, scale and time.[36] This includes working together to develop a culture of care, the valuing and sharing of care by challenging systemic injustices, protecting against precarity and vulnerability and long-term practices of collective care.[37] Within the con-text of action on climate change this in turn directs attention 'to the con-tent of care; to the ways in which care is marginalised; and to the need for ethical action'.[38] As suggested by Maria Puig de la Bellacassa,

> Thinking about and with care is compelling in this context because it offers possibilities for thinking commitment and obligation as non-normative forms of ethical engagement that could be more attuned to the decentring of human agency and privilege.[39]

Australian geographer Emma Power's notion of caring-with is instruc-tive here and focuses on the emergence of, and possibilities for, care in imperfect worlds. Caring-with involves three key frames of reference: the first situates care as a socio-material and performative agenda; the second places care in a temporal and socio-historical context; and the third recog-nises the care practices that take place and are negotiated across time, space and place.[40] Care-full practices of climate action as "caring-with" community emphasises questions surrounding why and how people take proactive interest in others, assume responsibility for their needs and take practical action to support their well-being.[41] These entangled relations of care arise from attunement to the needs of others (human and non-human) and their strengths/fragilities, as the guide for ethical action and relational capacity-building.[42]

Caring-with local places, people and spaces draws attention to new ways of conceptualising care: as always emergent and relational; as an assemblage of diverse materiality, actants, regulatory and policy frameworks, histories and socio-spatial contexts and so on; as a co-constituted public agenda rather than an individualistic pursuit; and involving the collective resources and networks necessary to sustain the capacity to care.[43] The importance of developing a culture of walking together with care seeks to support and cultivate proactive rather than reactive practices of collective community solidarity and care-full action to address the climate emergency. At the local scale this includes attentive observation and listening to others through which new sensitivities and capacities for caring-with are learned.

QUIET ADAPTATION

In this time of *the great dithering* (as the novelist Kim Stanley Robinson refers to the present age of global inaction on the climate crises), quiet climate action is vital to keeping things going along in politically fraught weather-worlds. Many climate advocates working in government and community settings intuitively and creatively work around the politics of climate consensus and of "fixing" climate action. A focus on everyday forms of sustainability is one way in which this work-around manifests. There are strong parallels between everyday sustainability—the work that people do in their homes, schools, communities and workplaces to address carbon emissions in energy, food, fashion, household items with the quieter forms of activism described by Laura Pottinger in her work on seed saving and gardening.

An extension of this is "quiet adaptation" which counters a tendency in climate planning and governance to seek community consensus around the urgency of climate change and the equally urgent need for climate action. Framed within a register of climate risks and emergencies, the need to form political and public consensus around climate change produces tensions between planning for certainty and planning for contingency. These tensions occur when competing demands to radically transform current maladaptive social norms and practices (the need to plan for contingency) are stymied or impeded by dominant economic, political and policy cycles, rules and legislation or by fragmented relationships between government, community and businesses (the need to plan for certainty).

In the recent Australian bushfire crisis, for example, an event that is still very much present in the hearts and minds of the authors and indeed many Australian people, one of the key activities of the subsequent Royal Commission into Natural Disaster Arrangements was to establish in partnership with the National Museum the 2019–2020 Bushfire History Project where people had the opportunity to share their photos, objects and experiences of fires. The Bushfire History Project represents a curated and more official version of community memory-work than that of the bushfire survivors' protest rallies in Canberra. In using the charred objects from the remains of their homes, the bushfire survivors' protests highlighted the different ways in which climate action and inaction are intimately connected and performed through the materials and infrastructures of our everyday lives.[44] In their very different ways both speak to the power of different, embodied and intimate ways of understanding, experiencing and acting on the localised impacts of climate change.

The common thread that connects the stories of "care-full" climate action in this chapter is the way in which the ordinary, felt and lived dimensions of climate change are entangled with "quieter" forms of climate adaptation. Across the different stories, the emphasis is on connection, experimentation and every day "labours of care" that work around-the-back of "technical-risk" approaches to mitigation and adaptation.[45] Indeed, approaches that build on what people care about and what they are actually doing together through embodied, everyday practices are vital to effective local-scale climate activism and adaptation. As one local government participant explained:

> there is interest and I just think probably what local government needs to do is to back away from the language of adaptation and move towards the things that the community already understand them as doing and build on that. Because we do see a lot of that. We see a very active community in plantings. When there are planting days, they are all out – rain, hail or shine. If there are dune [revegetation] days they are out rain, hail or shine. If there are clean up days, they are out there. If they do a consultation on the urban forestry strategy and biodiversity, they are all out there and in that consultation. Tiny houses – everyone is there.[46]

Quiet activism, in its attentiveness to embodied and creative forms of ordinary doing and working, inspires us to think about forms of quiet climate adaptation and action. For ordinary citizens, governments and community organisations, understanding what the risks are, how they are experienced, and who and what is in harm's way is a key critical action for bridging the environmental and social dimensions of climate change. But what quickly emerges is that people do not always understand or experience climate change in the same way. What is being described here is how working from practice, from experiments and with what people do and relate to is a powerful way of breaking down the real and perceived barriers to local climate adaptation.

'WE MAKE THE WEATHER'

In this chapter, we have focused on how local action on climate change can inspire different ways for people to 'walk together with care' across incredibly varied community, government and science domains. The local scale is where many people see and feel the effects of climate disruption and change. There are a myriad of pathways for working and walking together with care that "story" the quiet and diverse practices of local climate adaptation in Australian cities. While some of the projects are connected to wider local and regional government climate strategies (e.g. urban heat in Western Sydney), many of the projects work best from the ground-up. Importantly, this also promotes new pathways forward: where caring, sharing and co-benefits drive technical and economic innovations.

As outlined in the cases above, quiet climate actions—whether framed as climate art, sustainable household or community gardening practices, participatory street planting or community economy schemes to get solar panels on roofs or to start-up climate-friendly businesses—subtly and creatively subvert the boundaries between climate action and adaptation. By focusing on the intimate, the experiential and the everyday, quieter acts of climate action can show how different communities build repertoires of adaptive practice that are more flexible and open-ended, and which can communicate with each other in novel and creative ways.[47] Here, Mike Hulme's insight that 'we make the weather' and that the weather also 'makes us' is especially salient.[48]

However quiet activism is not without its limitations. We acknowledge that serious questions can be raised about the extent to which countless quiet and embodied acts of everyday action, subversion and subtle forms of resistance can be "scaled up". In the context of the climate emergency and crisis such as the Black Summer bushfires, 'walking together with care' might seem counterintuitive in its refusal to fix or resolve climate consensus. This can indeed seem a difficult proposition in the fiscally and risk-averse contexts that many people work in. But as this chapter has highlighted, rethinking the ways climate-activism can be done—*is* being done—and how these "walk together" to engage diverse and multiple collaborators through "other" modes, narratives and practices is a key social innovation. Equally serious is the question: what are the implications for individuals, communities, governments and the planet if working-with and working-around are the norm and where climate and biodiversity crises are not directly named or confronted?

The quiet proliferation of sustainable, regenerative and climate-adaptive actions at the local scale does not rely on fixing climate change or in establishing public or political consensus about climate change. In walking together with care, quiet activism offers valuable lessons as well as new ways of thinking about how people experience and act on climate risks and vulnerabilities. In the following chapter we turn to explicitly address the quiet capacity for climate activism at the local scale to be scaled out, up and deep as a care-full individual and community response to the climate emergency.

NOTES

1. Hulme, M. (2018) Weather-worlds of the Anthropocene and the End of Climate. Weber: *The Contemporary West, 34*(1), 59–70.
2. Head, L. and Gibson, C. (2012) Becoming Differently Modern: Geographic Contributions to a Generative Climate Politics. Progress in Human Geography 36(6): 699–714.
3. Climate for Change, accessed on https://www.climateforchange.org.au/
4. Highlighted in the Australian Research Council (ARC) DP150100299 interviews
5. Gibson-Graham, J.K. (2006*) A Postcapitalist Politics.* Minneapolis: University of Minnesota Press.
6. Ingold, T. (2010) Footprints through the weather-world: walking, breathing, knowing, in *Journal of the Royal Anthropological Institute, 16*(1), 121–139.
7. See also: Williams, M. (2017) Care-full Justice in the City. *Antipode*, 49: 821–839; Alam, A., and Houston D. (2020) Rethinking Care As Alternate Infrastructure, *Cities* 100: 1–10.

8. Tronto, J.C. (1993) *Moral boundaries: A political argument for an ethic of care.* Psychology Press.

9. Puig, de la Bellacasa, M. (2017) *Matters of Care: Speculative Ethics in More Than Human Worlds.* Minneapolis: University of Minnesota Press.

10. Ibid, p. 4.

11. Ibid, pp. 4–5.

12. Climate Council of Australia (2020) *Summer of Crisis Report.* Online: https://www.climatecouncil.org.au/wp-content/uploads/2020/03/Crisis-Summer-Report-200311.pdf [accessed 5 July 2020], p. 11

13. ABC News, 2019, *Grandmother dumps burnt remains of home at Parliament House in climate change protest* (December 3rd): https://www.abc.net.au/news/2019-12-02/bushfire-victim-nsw-nymboida-climate-change-protest/11757082

14. Blacktown City Council 2018, *Environmental Plans and Policies,* accessed 1 April 2019, https://www.blacktown.nsw.gov.au/About-Council/What-we-do/Environmental-Plans-and-Policies#section-2; Office of Environment and Heritage 2014, Metropolitan Sydney Climate change snapshot, NSW Government; Resilient Sydney 2018, Resilient Sydney: A strategy for city resilience, accessed 9 April 2019
https://www.cityofsydney.nsw.gov.au/__data/assets/pdf_file/0013/303700/Resilient-Sydney-A-strategy-for-city-resilience-2018.pdf; Western Sydney Regional Organisation of Councils 2018, Turn Down the Heat Strategy and Action Plan, accessed 9 April 2019 https://wsroc.com.au/media-a-resources/reports/send/3-reports/286-turn-down-the-heat-strategy-and-action-plan-2018

15. Ibid.

16. Abrahams, G., Johnson, B. & Gellatly, K. (2016) *Art+Climate=Change,* Melbourne University Press, Melbourne.

17. See more details in Karoly, D. (2015) *Climate science is looking to art to create change,* in The Conversation, May 7th, accessed on https://theconversation.com/climate-science-is-looking-to-art-to-create-change-41185

18. Lucas, C. H., Davison, A. 2019. Not 'getting on the bandwagon': When climate change is a matter of unconcern, in *Environment and planning E: Nature and space,* 2(1), pp. 129–149.

19. Anne Noble, *No Vertical Song,* accessed on https://ccp.org.au/exhibitions/all/no-vertical-song

20. One Planet, accessed on https://www.fremantle.wa.gov.au/OnePlanet

21. Bioregional, accessed on https://bioregional.com.au/

22. Fremantle, accessed on https://www.fremantle.wa.gov.au/sites/default/files/One%20Planet%20Communications%20Toolkit.pdf

23. One Planet Fremantle Strategy, accessed on https://www.fremantle.wa. gov.au/sites/default/files/sharepointdocs/One%20planet%20 Fremantle%20strategy-C-000307.pdf

24. One Planet Freomatch, accessed on https://www.fremantle.wa.gov.au/ one-planet/one-planet-freomatch; https://startsomegood.com/

25. Eco-eats, accessed on https://startsomegood.com/eco-eats; https:// startsomegood.com/disco-soup

26. Climate network, accessed on http://www.climatenetwork.org/profile/ member/climate-action-now-wingecarribee-can-win

27. Transition network, accessed on https://transitionnetwork.org/

28. Canwin climate action regional partners, accessed on https://corenafund. org.au/regional-partners/canwin-climate-action-now-wingecarribee-inc/

29. CaNWin, accessed on https://corenafund.org.au/

30. Ibid.

31. Moss Vale Community, accessed on http://mossvalecommunitygar-den.org.au/

32. Interview undertaken as part of our Australian Research Council (ARC) DP150100299 research.

33. Moss Vale Community garden video by Red Cup Films—https://www. youtube.com/watch?time_continue=404&v=Mee9_ydi4Tk

34. See Williams, M. J. (2020). The possibility of care-full cities. *Cities*, *98*, pp. 1–7.

35. Tronto, J. C., & Fisher, B. (1990). Toward a feminist theory of caring. In E. Abel & M. Nelson (Eds.), *Circles of care* (pp. 36–54). SUNY Press.

36. Tronto, J. (2017). There is an alternative: homines curans and the limits of neoliberalism. *International Journal of Care and Caring*, *1*(1), pp. 27–43.

37. Dowler, L., Cuomo, D., Ranjbar, A., Laliberte, N Christian, J. (2019) Care, in *keywords in radical geography: Antipode at 50 Antipode Foundation*, London, Wiley-Blackwell.

38. Lawson, V. (2009) Instead of radical geography, how about caring geography? *Antipode*, 41 (1) p. 210

39. Puig, de la Bellacasa, M. (2017) *Matters of Care: Speculative Ethics in More-than-Human Worlds*. Minneapolis and London: University of Minnesota Press.

40. Power, E. (2019) Assembling the capacity to care: Caring-with precarious housing *Transactions of the Institute of British Geographers* (2019), pp. 1–15, for application see also Hotker, M, Steele, W and Wiesel, I (2019) When gambling fails: Caring-with urban communities at the local scale, in *Cities*, 100, accessed on https://www.sciencedirect.com/sci-ence/article/pii/S0264275119314222

41. Conradson, D. (2003). Spaces of care in the city: the place of a community drop-in centre. *Social & Cultural Geography*, *4*(4), pp. 507–525.

42. Wiesel, I., Steele, W., & Houston, D. (2020). Cities of care: Introduction to a special issue. *Cities*, *105*, pp. 1–3.
43. Power (2019).
44. Alam, A. and Houston, D. (2020) Rethinking care as alternate infrastructure, *Cities*, 100, accessed online https://www.sciencedirect.com/science/article/abs/pii/S0264275119313484
45. Puig, de la Bellacasa (2017).
46. Alam and Houston (2020).
47. Houston, D., MacCallum, D., Steele, W., Byrne, J. (2016) Climate Cosmopolitics and the Possibilities of Urban Planning. *Nature and Culture* 11(3), pp. 259–277.
48. Hulme, M. (2018).

CHAPTER 5

Realising Transformative Potential

Abstract This chapter focuses on the potential of local activists to expand their reach: to transform practices and social relations and empower actors more widely ("scaling out"), more formally ("scaling up") and/or more profoundly ("scaling deep") than their original scope and scale. There are lots of different ways that these different kinds of scaling can happen. New practices change not just organisations, but people too. This is "scaling deep"—having a profound effect on the lives and minds of the people who participate in an initiative. It means that everyday activist practices can become embedded as a way of life—a new normal.

Keywords Transformation • Social innovation • Scaling • Everyday practices • Activism • New normal

New Normal/s

Adaptation to climate change is transformative: it involves significant changes to practices that are, in mainstream discourse, considered "normal". To adapt to climate change—a global, profound and (in human terms) potentially permanent crisis—we need "new normals" not just in pockets of local practice, but in society's institutions and structures. We need such fundamental changes because adaptation is going to include really hard stuff—dealing with the loss of settled coastlines, desolation of agricultural land, severe disruptions to industry, climate refugees, fossil

© The Author(s), under exclusive license to Springer Nature
Switzerland AG 2021
W. Steele et al., *Quiet Activism*,
https://doi.org/10.1007/978-3-030-78727-1_5

fuel dependence and unprecedented threats to biodiversity (to name a few). Our existing mainstream institutions are ill-equipped to deal with such challenges, not least because they support and are sustained by a competitive model of economic performance which requires that the economic utility of land and other natural resources be uninterrupted.

What practices need to be transformed to address the climate emergency? At the local scale, this often includes consumption practices—making the choice to consume less, or consume differently, in recognition that we may not be able to rely on such a steady supply of energy, water, goods and services in the future. It also includes production and trade practices, for instance, when people develop local networks for producing, preparing and exchanging food and/or other goods and services. In each of the examples outlined in the previous chapters—Climate for Change, Green Cross Australia, Environment House and the *ReNew* Initiatives, Gecko, Karl Mallon, Climarte, One Planet, CANWin and Mossvale Community garden—the emphasis was on identifying and implementing socially innovative responses for localised climate action.

These innovative, local-scale practices are not just economic; they are more broadly social. They are built on—and create relationships between—people and communities defined by shared interests, trust and solidarity. Indeed, without the support of such enabling relationships, many innovative ideas turn out to be pretty short lived or limited to small pockets of local activity. As outlined in Chap. 4, the diverse practices of climate action such as walking together and "caring-with" community emphasise the entangled relations of care that arise from attending to the needs of others (human and non-human). This is central to quiet activism as ethical action and relational capacity-building in response to the climate emergency at the local scale. As such, we see the transformative potential of adaptive innovation—even those that are facilitated by new technologies—as necessarily situated in the social realm.[1]

A second crucial aspect of successful activist practices is empowerment, which we understand as a process through which people gain the capacity to make change happen. This is both a precondition of socially driven change (for obvious reasons) and an important outcome of it, as participants in change-making practices gain collective resources—knowledge, social networks, infrastructures and so on—which can be further mobilised to institutionalise, disseminate and/or deepen the impact of new practices. The interaction of these two processes—relationship building and empowerment—can thus be remarkably powerful, leading to social, economic and political change beyond the ambit of the initial action. Ultimately, as new practices and relations become embedded across time/ space and scale they have potential to become the new normal. This emphasis on the horizontal and vertical dimensions of making and breaking connections will be discussed further in relation to socially innovative practices and the role of landscapes, regimes and niches in the following chapter.

This chapter builds on the key themes of quiet activism outlined in the previous chapters—that is, building and bridging the knowledge base (Chap. 2), bringing missing actors to the table (Chap. 3) and walking together with care (Chap. 4)—to explore in more detail some of the processes by which local-scale, "quiet" social innovations take hold and expand their reach, shaping new normals well beyond their original sphere of influence. The focus is the realisation of transformative processes specifically in relation to climate action at the local scale which is complex: involving new relational practices between human, technological and other non-human actors (animals, plants, minerals, rivers, mountains, the weather, etc.).

We start this discussion by introducing the transformative typology of scaling out, up and deep, and then go on to describe the two case studies from our research—Solar $aver (in Melbourne) and Ecoburbia (in Perth). The final sections of the chapter take a different approach to the previous chapters to explore in more detail the three forms of scaling and the ways in which both cases reflect all scales but do so in very different ways: one being a government-led and the other a grass-roots project.[2] In these two key cases, the highlighted quiet activist practices can become embedded as a way of life—a new normal.

Scaling Out, Up and Deep

Our argument outlined throughout this book is that quiet activism, undertaken at the local scale, is as important to the transformation of society as more visible and familiar forms of activism such as lobbying and mass protest. In this chapter we extend a typology of scaling originally developed and described by social innovation scholars Frances Westley and Nino Antadze,[3] which differentiates between *scaling out*—as the replication of practices in different places—and *scaling up*—which is when practices become institutionalised as, for example, policy or new procedural models. This typology was further developed by feminist scholar

Michele-Lee Moore and colleagues through the addition of the idea of *scaling deep*, which challenges the "size matters'" assumption of conventional innovation analysis by positing that internalisation is also important: innovative practices can work, and can become transformative, by making a profound difference to the lives, minds, relations and feelings of those who participate.[4]

Scaling is enabled by a range of processes. Different types of actors come into contact through different types of technologies, using different systems and modes of communication, and responding to opportunities that resonate with their different concerns and daily practices. Central to *scaling out* are various ways which allow actors not associated with the original actions to adapt and replicate them in other places. These different modes of influence are important because simple replication often fails because local initiatives are fundamentally shaped by their environmental, social and (especially) political contexts. However, this does not address the higher level institutionalisation of innovation that is concomitantly needed by way of new or reinterpreted policy and public governance arrangements. The processes of *scaling up* are therefore also critically significant as a means of generating changes to the political context in ways that allow local experiments to sustain themselves and to flourish.

Scaling up involves new kinds of cooperation between local and state actors in which policy and practice are neither top-down nor bottom-up but emerge from the cooperation itself. This involves cooperation, and the infrastructure that facilitates it, needed to empower both parties, opening new possibilities for achieving change together. This also involves finding a way of talking about climate change and climate action that resonates with the concerns and priorities of both parties—that is hopeful rather than despairing—to help shift entrenched ways of thinking about what is, and isn't, possible. Finally, new practices can change not just organisations, but people too. This is *scaling deep* which builds on scaling up and out to involve a profound effect on the lives and minds of the people who participate.

SOLAR $AVER

Rising energy needs, combined with the rising unit cost of electricity, is a "here and now" issue with which consumers can readily engage. Its financial aspect connects it clearly with socio-economic equity, compelling sustainability officers to consider the distributional impacts of their responses.

Instead of being a passive actor in the face of climate change, the local government of Darebin in Victoria, Australia, is taking actions that are intended to have a transformative effect on the way that people in their city view energy and its effect on the environment.

The *Darebin Climate Emergency Plan* 2017–2022 was adopted in 2017. It is an explicit reframing of the discourse around climate change from that of "risk", with its connotations of uncertainty and technical analysis, to that of "emergency", implying the need for urgent and radical action right now. This plan argues that to avoid dire consequences, we need to enact a complete transformation of the way that energy is considered, produced and used.

> We recognise that we are in a state of climate emergency, and we urgently need to take action to avoid dangerous climate change and provide maximum protection for people and nature.[5]

The City has implemented a number of strategies to these ends. Examples include retrofitting City infrastructure to improve energy efficiency, providing households with window shades and weather proofing to reduce energy costs, helping business to switch to LED light bulbs through their *Light$mart* programme, investing with fossil fuel free financial institutions where possible, providing incentives to staff to use environmentally friendly transport options and the organisation of a climate change emergency conference. The input and involvement of Darebin's diverse community has been a central component of efforts to change the City's approach to energy, with consultation undertaken at various stages of projects to ensure that they are culturally and socially appropriate.

One of the most successful and revolutionary programmes that Darebin has established is the Solar $aver programme, which synergistically responds to several related concerns: managing health risks associated with heatwaves, transitioning to renewable energy and distributional justice. The programme responds to concerns within the community around the management of health risks associated with heatwaves, especially for elderly citizens, avoidance of maladaptive responses, and the transition to renewable energy and distributional justice in the face of climate change.

> We're comfortable now with saying this is not a maladaptation strategy. This is actually a complementary strategy.

For example, when we're developing the Solar Saver program – Councillors say we're not interested in offering this to the community, this is a special deal and we'll offer it to disadvantaged households – pensioners – that's a target group …

That sits pretty comfortably with us. In my life I have worked in the social sector as well with long term unemployed and refugee communities etc so from a personal point of view it ticks some boxes for me and I'm really passionate and love doing it.[6]

As outlined earlier, Solar $aver was deemed the most responsible way that the City could help those who were struggling economically to access the energy necessary to cope with extreme temperatures without contributing further to greenhouse gas emissions—that is, it is an adaptive response that is not maladaptive. It was a significant departure from traditional support programmes that provided subsidies and discounts to improve the affordability of renewable energy infrastructure but failed to benefit those people and communities who could not afford the upfront costs.

A survey of pensioner households identified that 73% could not afford the upfront cost of solar panels and, more surprisingly, that many were also unable to afford the cost of their air-conditioning on hot days. In response, the City developed an arrangement for financing solar panels on interest-free terms, recouping costs over ten years through a special charge in addition to annual rates, calculated to be lower than residents' estimated power bills in the absence of the solar panels. The programme was initially restricted to pensioners, but having proven highly successful in terms of both take-up and cost recovery, it has been gradually extended to other households, social housing providers, community groups and businesses. This social innovation has been achieved through a significant rethinking of the relationship between the local government and its citizens, in that the City has become, for those in need, a financial service provider, as well as a provider of more traditional services.

An important side effect of this programme is that owning solar panels and a commitment to renewable energy have become a matter of pride (and envy) in Darebin's pensioner community, and this plays a role in building political support for a reconfigured state energy regime that privileges renewables over fossil fuels.[7] It has also received strong publicity, and prompted the development of similar local government initiatives in Victoria and beyond, helping to refocus attention on the equity implications of climate change, and climate change responses, at scales below the global.

Whilst the Solar $aver programme's focus on socio-economic equity remains an exception, even in Darebin, it shows that it is possible to respond sensitively and effectively to uneven vulnerabilities at the municipal level, albeit without seriously challenging neoliberal discourses of individualism or market-led growth. We hope that this may become a point of departure for environmental justice to play a greater role in a wider range of adaptation measures—although at the time of writing we found little evidence that these considerations are shaping other areas of adaptive activity within local governments. For example, urban forest and green infrastructure strategies—consistently documented as a means of managing heat—do not generally appear to be targeted to poor areas, in spite of these areas often being poorly serviced by parks.[8]

Ecoburbia

Ecoburbia is, formally, a small business in Perth, Western Australia, run by Shani Graham and Tim Darby, but it is also the realisation of a journey towards increased resilience and sustainability, both on a personal level and on a community level. Shani and Tim have established several projects which respond to their concerns about peak oil production, climate change, economic instability and environmental issues, while creating a life that helped them to overcome physical and mental exhaustion. Shani and Tim's endeavour has gradually transformed from providing sustainable tourist accommodation in 2005 to educating the community through practice and challenging traditional constructions of suburban life in the hope of creating something more resilient.

> Ecoburbia is part urban infill development, where we have converted one house into four self-contained living units, tripling the population density without adding to the house's footprint.
>
> Ecoburbia is part urban farm with chickens, goats, compost, fruit trees, plus a large shared veggie patch.
>
> Ecoburbia is part demonstration sustainable house, with cutting edge energy systems, water collection and dispersal systems and innovative passive solar design.
>
> Ecoburbia is part educational opportunity and community hub, with regular tours, workshops, films and other community events.[9]

In 2013, Shani and Tim purchased a house on a quarter acre block in a middle suburb of Perth, which would become not only the home, but also the key achievement of Ecoburbia. They describe what they have created as an 'alternative urban infill development, housing cooperative, community garden, model sustainable house and benevolent dictatorship'.[10] In a time of population growth and increased urban density, Ecoburbia demonstrates that people can live in smaller, less consumptive spaces without sacrificing the benefits of open space and productive land.

The house has been divided into four and a half separate living spaces and a number of communal areas, with up to ten people of varying ages calling it home. In contrast to conventional forms of subdivision, these changes have not altered the building's footprint, leaving space for a 350-square-metre garden in which produce is grown and both chickens and goats are raised (see Fig. 5.1). This communal garden, open to neighbours who often bring scraps for the animals, is Shani and Tim's way of demonstrating that simple changes can be adopted to increase resource independence.

Additionally, there are four water sources on the property including a bore and a 50,000-litre underground tank, grey water is recycled in the garden, solar systems have been installed, the building has been retrofitted to make it solar passive and chemical cleaning agents are avoided. All these measures reflect Shani and Tim's determination to move towards a "new normal", in which resource instability is combated by households and communities empowering themselves through direct action.

A central element of Ecoburbia is its role as an educational space, helping to spread sustainable thinking and practices to all sectors of the community, enabling them to make changes that will mitigate the risks and effects of climate change, peak oil and economic instability. There is a clear focus on positivity in Tim and Shani's approach to education, choosing not to weigh people down with the overwhelming nature of climate change, instead endeavouring to instil agency in those who attend Ecoburbia, helping to develop knowledge that will enable people to make their own individual difference.

Shani and Tim deliver their message in a number of ways, including as facilitators of Perth's very successful "Living Smart" programme as well as by hosting seminars from sustainability experts, holding movie and discussion sessions, tours and giving talks to interested groups (Fig. 5.2).

As Shani and Tim describe it, this includes:

Fig. 5.1 Ecoburbia urban farm cooperative. (Source: Ecoburbia)

- Opening our home to the general public for tours and short courses
- Presenting long and short courses outside Ecoburbia
- Hosting speakers and workshops at Ecoburbia
- Providing an affordable venue for other sustainability activities
- Celebrating being a sustainable community by running community events
- Being a conduit for change linking individuals and groups interested in sustainability
- Providing support and sponsorship to other local sustainable projects.

Through the nurturing of relationships in their local community and in order to realise transformative potential, Shani and Tim are demonstrating a style of resource consumption that minimises reliance on mainstream commercial suppliers in favour of sustainable sources that foster resilience and a sense of individual empowerment.

Ecoburbia is a home. It's a beautiful home, with a beautiful garden, full of creativity and love. It's a home that we share – with those who live here, with those who visit. It is a safe place – we grow food, share our space with productive animals, make power, collect water. Sharing this with others gives us hope for the future.[11]

Fig. 5.2 Ecoburbia sustainability thinking and practices in action. (Source: Ecoburbia)

SCALING OUT

As the stories linked to this theme show, there are lots of different ways that scaling can work. Networking activities through which knowledge and practices can be exchanged and built (see Chap. 2) are important for scaling out, and our interviews suggested that the more diverse these activities are the better. Different types of actors come into contact through different types of technologies, using different systems and modes of communication, and responding to opportunities that resonate with their different concerns and daily practices (see Chaps. 3 and 4).

As an example, Darebin Council are active participants in established networks such as the *Cities Power Partnership* and the Global Covenant of Mayors for Climate and Energy. But they are also part of a novel type of cooperation among a group of neighbouring local governments, the *Northern Alliance for Greenhouse Action*. Rather than collaborating on discrete regional projects, as is common,[12] officers address their efforts to sustainability "problems" (e.g. becoming carbon neutral, preparing for heatwaves, adaptive gardening) as local opportunities arise, and they agree to share their experiences and successful models.

> Yeah – that's the … Alliance. And that rivalry is used to help stretch us. They might work on something that we haven't worked on; we might work on something that they haven't worked on. We borrow from each other.[13]

This gives them a degree of flexibility for innovation, as they are not required to conform to top-down resourcing agreements between their councils, produce a predetermined outcome (such as a regional adaptation plan) or work to a compromised timeframe. At the same time, it reduces the corporate risks of wholesale experimentation by grounding new projects in lessons learnt in relatively similar contexts to their own. Thus, this 'greenhouse alliance'—one of several in Victoria—is seen as an effective strategy for out-scaling successful local government innovations, such as Solar $aver, to the regional level and beyond.

Also, important to scaling out are various modes of influence which encourage actors not associated with the original actions to adopt and adapt them in other places. Such modes can be direct and formal (teaching, training, consultancy, publication, etc.) or informal and less direct (modelling by example, engaging in diverse forms of public discourse, etc.). In Ecoburbia's case, their educational and awareness-raising activities are well attended by households from across the metropolitan region, and they offer tailored training programmes for public and private sector organisations. Clever use of information technology has also helped to make their skills and philosophy accessible on a global scale and allowed their business to adapt to the limitations posed by the COVID-19 pandemic.

A freely accessible website, blog and Facebook page provide updates on their activities and adaptable tips for living a more sustainable life. The sharing of their story has strongly influenced other local activities throughout Perth, interstate and in North America, and they note that their approach is broadening its appeal across sectors, including among government and corporate entities. A 2018 segment on a national television show and invited speaking tour of Canada demonstrates the growing interest in Ecoburbia and their innovative practices around the world.

In a different vein, Darebin's Solar $aver provides an adaptable model for financing innovative programmes, which has been widely disseminated and adopted by other Councils in Victoria and other states. Significant media coverage and Darebin's winning of the environmental justice category in the 2015 Premier's Sustainability Awards have increased their influence and assisted these out-scaling processes.

Scaling Up

Important as such opportunities for transfer and replication are, we do not believe that they are sufficient for realising the transformative potential of local initiatives. Indeed, simple replication often fails because local initiatives are fundamentally shaped by their environmental, social and (especially) political contexts. This is why we place importance on processes of *scaling up*, meaning the higher level institutionalisation of innovation by way of new or reinterpreted policy and public governance arrangements—that is, changes to the political context in ways that allow local experiments to sustain themselves and to flourish. In Darebin, scaling up is key to Solar $aver's success: the development of a formal mechanism including newly configured partnerships between Council, solar infrastructure providers and citizens, are necessary precursors to the programme's roll out.

Scaling up involves new kinds of cooperation between local and state actors in which policy and practice are neither top-down nor bottom-up, but emerge from the cooperation itself—what we, following colleagues in Europe,[14] have described as bottom-linked governance. There are several aspects to this. One, which many of the people we spoke to alluded to, is finding a way of talking about climate change and climate action that is hopeful rather than despairing, that resonates with the concerns and priorities of both parties, and that helps shift entrenched ways of thinking about what is, and isn't, possible.[15] Another is that the cooperation and the infrastructure that facilitates it need to empower both parties, opening new possibilities for achieving change together—we can see this at work particularly in the Ecoburbia story.

In keeping with their philosophy of small-scale change, Shani and Tim avoid direct involvement in politics. However, we see Ecoburbia as profoundly political in both its drivers and its effects. Firstly, it builds community at and beyond the local scale—directly through its activities to share resources, skills and knowledge, and also indirectly through its existential challenges to conventional urban living, which constantly generate and practically resolve tensions between residents, neighbours and the City. As one example among many, the legal acquisition of two milking goats, Pumpkin and Whimsy, required negotiation to reach an agreement that reassured neighbours that they would not be adversely affected. After this, Pumpkin and Whimsy became popular residents of the neighbourhood, whose presence brought people together and generated sustained interest in, and conversation about, possibilities for alternative ways of living in the suburbs.

Secondly, Ecoburbia has been created not just as a community resource, but as a registered business that demonstrates on the one hand the capacity for sustainable living to be financially viable and, on the other, an alternative model of economic sustainability that is not dependent on continuous growth. Shani and Tim describe the profitability of their ventures as a peripheral matter, preferring to reduce their financial requirements rather than expand the business. Thus, Ecoburbia disrupts accepted competitive economic regimes through the adoption of practices grounded in relations of respect and solidarity, not only with people, but also with the environment. The alternative they offer is recognised institutionally as a valid form of private enterprise.

Thirdly, the cultivation of a constructive, open relationship with Ecoburbia's local council has not only enabled the pragmatic resolution of differences before conflict arises, but also allowed their ideas and approaches to be sympathetically heard and promoted in Council's own policy and plans, for example in relation to community gardening and street parties.

Scaling Deep

In conventional terms, we might understand bottom-linked governance as a capacity to find and deliver "win-win" actions. Win-win can, however, be a double-edged sword, especially for government agencies working within a broader policy framework which, in general, does not support actions that seriously challenge or undermine the principles of growth capitalism.[16] But sometimes something else can happen: new practices change not just organisations, but people too. Scaling deep matters. It means that innovative practices can become embedded as a way of life—a new normal. And new normals generate new communities of interest, which can become political forces with potential to challenge the win-win trap.

For instance, local government actions are generally constrained by neoliberal discourses of individualism, property ownership and market-led growth, and the Solar $aver programme has been implemented within these constraints. But by sensitively targeting the inequities in climate vulnerability, Darebin has helped to quietly mobilise a hitherto relatively marginalised community. An important side effect of the Solar $aver programme is that owning solar panels has become a matter of pride (and envy) among Darebin's pensioner community and beyond. This is playing a role in building political support for a reconfigured state energy regime that privileges renewables over fossil fuels.[17]

Turning to Ecoburbia, whose protagonists are relatively free from the formal political constraints faced by local government, scaling deep has been arguably their most powerful transformative strategy. For Ecoburbia's participants, living in an environmentally and socially conscious manner has become part of a new normal, a different way of thinking about the possibilities for a more comfortable and resilient future. An ex-resident commented on how the experience of living there had profoundly changed her day-to-day expectations and practices, not only in terms of the way she, as an individual, consumed resources, but also as a citizen and a neighbour. This is, if you like, a "quiet" social movement, empowering a growing community with the conceptual and practical means to view economic well-being in terms other than profit/growth, to act on their own felt vulnerabilities without the need for any institutional mandate,[18] and to find shared concerns as a basis for action in the face of social diversity.

Quiet Activism: Means *and* Ends

As a collection of socially innovative practices undertaken at the local scale, quiet activism does not separate means from material ends. Rather, it responds to needs (material, cultural, ecological, etc.) unmet by state and market through the adoption of new/experimental social practices[19] and informal and institutional forms[20] that provide for democratic and solidarity-based processes. Socially innovative strategies thus can lead to the collective empowerment of a range of diverse community actors,[21] and ideally this can create the social conditions for their own expansion and for broader and/or deeper social change that more actively includes and addresses the needs of those most marginalised in the communities of quiet climate activism and practice at the local scale.

Quiet activist strategies evident at the grass-roots level of the cases outlined in this chapter focus on all three levels of scaling—scaling out, up and deep—as part of reorienting everyday practices from present to future conditions. This is a prefigurative politics—starting now to live in the future, for example, by attempting to localise production-consumption networks as far as possible, or planting tree species that will be adapted to future climate shifts in temperature and rainfall. An important aspect of this strategy is that it builds on local sense of place and frees people to act at their own scale of influence, a scale defined relationally, rather than in terms of cartographic boundaries.[22]

> I've been telling people coming in for the native plants that this is a better plant – a drier climate plant. ... So those plants more north of Perth ... those sorts of plants are more suited to our climate because it's drying. So the drier climate plants are becoming a bit more prevalent.
>
> Some very passionate people, a lot of time and energy, thinking about local trading systems, local food production, getting the messages out there—they make a marvellous effort.
>
> I think one of the things that happens ... you know when I said not everything upscales? Well not everything replicates. Like, most things that are successful have the basis of success is relationships. I think that's one of the things that people do is go, 'Well that works. Let's pick that up and put it somewhere'. And the thing that they primarily miss is the relationships of trust and the connections with community that actually are the things that make that work.[23]

This idea is an important driver for many participants in Transition Towns and similar organisations which strive to turn the anticipation of loss into positive action.[24] It connects practices in local space-time with grander scales through an underlying theory of transition, whereby transformative ideas and practices develop in experimental niches, eventually evolving through diffusion into systemic change[25] or, in the case of climate change and peak oil, growing to fill the gap when the existing systemic regime collapses.[26]

Our interviews with local actors in cases across Australia highlighted a number of tensions and paradoxes related to the position of climate change as a highly political issue within a depoliticised policy environment. Frustrating as these contradictions are, participants were often able use them generatively, as lines of flight for locally relevant practices that connect climate change to felt human needs. These initiatives are, to varying extents, socially innovative in that they seek to rearticulate institutional relations and work to build collective capacity around issues of local concern, with some showing emerging potential for enhancing social inclusion and distributional justice.

However, while such initiatives appear to diffuse easily (scaling out), their transformative potential (scaling up) is less evident. Constrained by a political consensus that places economic growth at the centre of institutional mandates, local governments in particular are limited to initiatives which can demonstrate success in short time frames without harming the capitalist status quo. For example, while risk management frameworks enable the identification of vulnerabilities and provide an institutional site for a number of adaptive strategies, they are also prone to misrecognition. The reductionist processes they require tend to overlook the entanglements between risk, space-time, culture, politics and inequity. This can make what we might call situated risks (i.e. being unemployed, Indigenous, food insecure, etc.) difficult to discern and almost impossible to institute in a holistic way into corporate "business as usual".

It can be difficult to see such activities as transformative in the context of the climate emergency—they appear to do nothing to reframe discourses of market-driven growth, they largely support a competitive model of urban development,[27] and they can be critiqued as a shift in responsibilities and costs from the state to civil society. However, our interview participants and our own critical engagement with the local, everyday practices of quiet activism convinced us they are vitally important. In particular, they provide a means of engaging citizens who might otherwise be unlikely to get involved with climate action, or to build and maintain relationships with their community and with each other. As such, they help to enrol a polity concerned with the effects of climate change, if not with climate change itself, potentially overcoming impasses that might otherwise prevent action.

Our stories of quiet activism highlight the importance not only of scaling out and up, but also of scaling deep in creating these conditions—because "new normals" are not simply a matter of institutional discourse; they are realised in the concrete social-economic-ecological-political practices of people and communities, living their lives. We join here with those seeking an expanded and relational notion of connection to places, people and species which is needed in the face of climate crisis. As outlined in the Shadow Places Manifesto, 'the forces and effects of this crisis are simultaneously global and local, connecting places across vast distances through fossil fuel economies, and with devastating consequences for places, people and species, permeating all Earth's life-support systems'.[28]

Key to the transformative potential of social innovation is the capacity to reframe hegemonic discourses and practices.[29] This means focusing not only on "win-win" solutions—though these surely have their place—but also on actions that challenge the underlying assumptions that make win-win important—that is, on political transformation. This is a theme taken up and extended in our next chapter focused on making and breaking connections and the varied ways in which local innovative activist practices connect with other practices—horizontally and vertically—to bring about change at the local scale in response to the climate crisis.

NOTES

1. Moulaert, F., MacCallum, D., Mehmood, A. and Hamdouch, A. (eds) (2013) *International Handbook on Social Innovation*, Elgar, Cheltenham.
2. We are avoiding the language of "top-down" versus "bottom-up". Successful social innovations can rarely be strictly categorised as one or the other; they are usually better described as "bottom-linked" (see Moulaert et al. 2019).
3. Westley, F.R. and Antadze, N. (2010), 'Making a Difference: Strategies for Scaling Social Innovation for Greater Impact', *The Innovation Journal: The Public Sector Innovation Journal* 15 (2) http://innovation.cc/scholarly-style/westley2antadze2make_difference_final.pdf
4. Moore, M., Riddell, D. and Vocisano, D. (2015), 'Scaling our, scaling up, scaling deep. *Social Innovation* 58: 67–84.
5. Darebin Council Climate Emergency, accessed on http://www.darebin.vic.gov.au/en/Your-Say/Advocacy/Climate-Emergency
6. Excerpt from an interview undertaken as part of the Australian Research Council (ARC) DP150100299 project transcripts.
7. Slezak, M. (2017) Renewables roadshow: how the 'nonna effect' got Darebin's pensioners signing up to solar, *The Guardian*, 22 March. https://www.theguardian.com/environment/2017/mar/22/renewables-roadshow-how-the-nonna-effect-changed-darebins-approach-to-solar
8. Ambrey, C., J. Byrne, J., T. Matthews, A. Davison, C. Portanger and Lo, A. (2017) Cultivating climate justice: Green infrastructure and suburban disadvantage in Australia, *Applied Geography* 89, pp. 52–60.
9. See Ecoburbia, accessed on https://ecoburbia.com.au/
10. Ibid.
11. Ibid.
12. See Schlosberg, D., L.B. Collins and Niemeyer, S. (2017). Adaptation policy and community discourse: risk, vulnerability, and just transformation. *Environmental Politics*, 26(3), 413–37.

13. An interview undertaken as part of the Australian Research Council (ARC) DP150100299 project transcripts.

14. This term was coined by Marisol Garcia and colleagues at the University of Barcelona (refs). It has become common parlance in work about social innovation in local and regional development (see Moulaert et al. 2019).

15. "Reframing the discourse", a characteristic of democratic leadership—see Parés, M., Ospina, S.M. and Subirats, J. (2017), *Social Innovation and Democratic Leadership: Communities and Social Change from Below*, Elgar, Cheltenham.

16. This insight has led some scholars to talk about a "post-political" era—see for instance Wilson, J. and Swyngedouw, E. (eds) (2015) *The Post-Political and its Discontents: Spaces of Depoliticisation*; Spectres of Radical Politics, Edinburgh University Press.

17. Slezak, M. (2017) Renewables roadshow: how the 'nonna effect' got Darebin's pensioners signing up to solar, *The Guardian*, 22 March. https://www.theguardian.com/environment/2017/mar/22/renewables-roadshow-how-the-nonna-effect-changed-darebins-approach-to-solar

18. See also Houston, D., MacCallum, D. Steele, W. and Byrne, J. 2016. Climate cosmopolitics and the possibilities for urban planning. *Nature and Culture* 11 (3): 259–277.

19. Howaldt, J. and M. Schwarz (2016), 'Verifying existing Social Theories in reference to Social Innovation and its Relationship to Social Change', SI-DRIVE Deliverable 1.3 https://www.si-drive.eu/wp-content/uploads/2016/07/SI-DRIVE-D1-3-Social-Change-final-260416-2.pdf

20. Garcia, M., and S. Vicari Haddock (2016). Special issue: housing and community needs and social innovation responses in times of crisis. *Journal of Housing and the Built Environment* 31(3), 393–407.

21. Moulaert, F., D. MacCallum and J. Hillier (2013) Social Innovation: intuition, precept, concept, theory and practice", in F. Moulaert et al. (eds.) *International Handbook of Social Innovation,* Elgar, Cheltenham.

22. cf. Scannell, L. and R. Gifford (2013) Personally relevant climate change: The role of place attachment and local versus global message framing in engagement. *Environment and Behavior* 45(1): 60–85; Agyeman, J., D. Schlosberg, L. Craven, and C. Matthews (2016) Trends and directions in environmental justice: from inequity to everyday life, community, and just sustainabilities. *Annual Review of Environment and Resources* 41, 321–40.

23. Excerpts from three interviews undertaken as part of the Australian Research Council (ARC) DP150100299 project transcripts.

24. Head, L. (2016) *Hope and Grief in the Anthropocene,* Routledge, London.

25. See Kemp, R., D. Loorbach and J. Rotmans (2007) Transition management as a model for managing processes of co-evolution towards sustainable development. *The International Journal of Sustainable Development & World* Ecology 14(1), 78–91; Bulkeley, H., and M. Betsill (2005). Rethinking sustainable cities: Multilevel governance and the 'urban' politics of climate change. *Environmental politics* 14(1), 42–63.

26. Seyfang, G. and A. Haxeltine (2012) Growing grassroots innovations: exploring the role of community-based initiatives in governing sustainable energy transitions. *Environment and Planning C: Government and Policy* 30, pp. 381–400.

27. See Kenis, A. and M. Lievens (2017) Imagining the carbon neutral city: The (post) politics of time and space. *Environment and Planning A* 49(8), 1762–1778; Geels, F.W. (2014) Regime Resistance against Low-Carbon Transitions: Introducing Politics and Power into the Multi-Level Perspective. *Theory, Culture and Society* 31(5), 21–40.

Potter, E., F. P. Miller, E. Lovbrand, D. Houston, J. McLean, E. O'Gorman, C. Evers and G. Ziervogel (2020) A manifesto for shadow places: Re-imagining and co-producing connections for justice in an era of climate change. Nature and Space, accessed on https://doi.org/10.1177/2514848620977022

28. Parés, M., S.M. Ospina and J. Subirats (2017), *Social Innovation and Democratic Leadership: Communities and Social Change from Below,* Elgar, Northampton.

CHAPTER 6

Making and Breaking Connections

Abstract In this chapter the focus is on innovative activist practices (by local government authorities, non-government organisations and the private sector) and how they connect with other practices. In doing so we identify the critical points of connection and the potential footholds or leverage points for transformational change in response to the climate crisis. Quiet activist practices are made up of many elements which can be grouped together as images or meanings, materials and skills or competences. Making and breaking connections represents provocative challenges to the status quo. These activist practices can be creative disruptors in order to engender transformative potential and politicised change to address the climate emergency.

Keywords Practices • Connections • Multi-scalar perspective • Niches • Regimes • Landscapes

QUIET INNOVATION AND ACTIVIST PRACTICES

Change or the transformation of activist practices—quiet or otherwise—requires a multilevel perspective which highlights the connections, or points of intersection, between scale and praxis. The previous chapter explored the role of quiet activism for realising transformative potential by scaling out, up and deep at the local scale in response to the climate crisis.

In this chapter we develop the fifth and final key theme in the Quiet Activist Framework focused on making and breaking connections. Our focus here is how and in what ways local innovative activist practices connect with other practices to bring about change.

As we emphasise throughout this book, social change does not happen just at a macro-scale. Individuals, collectives and their relationships are important innovators of new ways of thinking and acting. Quiet activism is enacted through developing interpersonal relations where 'a smile, a nod, a mutter of encouragement, a brief conversation over a counter' may be significant means of building relationality.[1] Such acts would traditionally be considered too insignificant to count as activism. They may not, in themselves, generate transformation, but may well foster relationships which lead to future change.

Change in the form of local action on climate change may be facilitated by the circulation of ideas and adjustment of practices across a horizontal plane. As Elizabeth Shove points out, 'from a policy perspective, theoretical [and empirical] accounts of socio-technical regimes and landscapes [and niches] are of limited value unless they can be used to identify points at which the cause of path dependent development might be turned in a more sustainable direction'.[2] Shove outlines a model which vertically distinguishes three "planes": a macro-scale socio-technical landscape, a meso-scale patchwork of regimes and a micro-scale of "niches" where novel ideas, processes and products are developed.

We develop this model into a multiscale perspective of practices and connections for understanding quiet activism as socially innovative practices in Fig. 6.1, to draw attention to both the *vertical relationships* between landscapes, regimes and niches and *horizontal elements* of practice, and to the points of intersection between practices and regimes in the social innovation processes at the local scale.

We cannot overemphasise the importance of making new connections for quiet activism—the "fitting together" of symbolically and technically coherent practices[3]—for disseminating ideas and practices and also for breaking old connections in terms of disrupting unsustainable, outdated ways of thinking and acting. Understanding how actors, ideas and artefacts become embedded in or dislodged from ordinary practices might reveal points of leverage for change.[4] Actors can employ three strategies to influence change: voicing ideas and convincing people in technological and institutional frameworks; shaping technological nexus by making or breaking/disrupting connections between ideas, practices and their

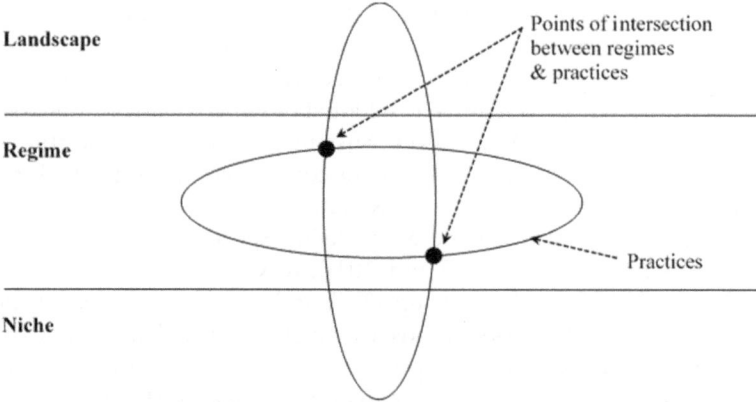

Fig. 6.1 A multi-scale perspective of connections and practices. (Source: Hillier 2021 adapted from: Geels (2011); Hargreaves et al. (2013); Rip and Kemp (1998); Shove (2003)

institutional carriers; and creating niches of new ways of thinking and acting and building a constituency behind them.[5]

An emphasis on relational active negotiation and persuasion characterises practice theory. As Wetherell et al. explain, a practice approach 'is interested in how people make sense of their circumstances and negotiate and initiate patterns of activity in concert with others'.[6] Practices are made up of many elements comprising *images* (meanings, symbols), *skills* (know-how, forms of competence) and *materials* (artefacts, technologies) which are actively and recursively integrated through everyday performance.[7] We can take the example of people attempting to be comfortable in the heat of summer. The image may be of coolness, meaning the ability to sleep at night and wellbeing. Materials include air conditioners, fans, light clothing, a home with shade, windows, eaves, trees and probably a mortgage or owning one's home. Skills/competences include questions of can I afford to use the air conditioner or fans? What can I do to cool my home and myself? Do I have appropriate clothing, shade and so on? Every time someone turns on the air conditioner at home, they combine images (of coolness), skill (knowing how to operate the machine, closing windows) and materials (the air conditioner, electricity system, windows and so on).

This chapter highlights four cases by local government authorities, non-government organisations and private sector individuals to identify the critical points of connection and the potential footholds or leverage points for transformational change in response to the climate crisis. All four cases depend on a crucial role played by individuals in communities. From Griff Morris, an individual residential designer-builder embedded in the Housing Industry Association, Sustainable Energy Association and Australian Commonwealth Government committees, to individuals at the former Marrickville Council and Renew (the Alternative Technology Association) who organised the Marrickville Speed Date a Sustainability Expert (SDSE) events in 2015 and 2016, to Geoffrey Love establishing the Elwood Floods Action Group (EFLAG), a local citizens' group in response to the 2011 Elwood floods, to the officer at Redland City Council who persuaded colleagues and political members to form RedWaste, a Council Business Unit engaged in recycling solid waste diverted from landfill.

If we look at innovative practices (by local government authorities, non-government organisations and the private sector) and how they connect with other practices, we can begin to identify the critical points of connection and the potential footholds or leverage points for change. It is these points which are emphasised in Figs. 6.2, 6.3, 6.4 and 6.5.[8] The four selected cases illustrate how bottom-up, bottom-linked and top-down socially innovative, climate-change-related activities can make a difference through disrupting and transforming practices. Whether the indirect action of quiet persuasion or more direct action of campaigning and lobbying, changes to practice regimes have occurred. New connections have been established and some old connections broken as small, local acts of making and doing 'critique, subvert and rework'[9] previously dominant modes.

This is no quick fix or panacea. As the cases also demonstrate, not all social innovations are adopted successfully. There are always barriers to their introduction and operation. For instance, economic barriers (the innovation is unable to compete in the market given regulatory and other constraints), technical barriers (a lack of skills to operationalise the innovation), social and institutional (inertia of existing practices and perceptions may be difficult to overcome). Nevertheless, we offer the four cases below as examples of local, socially innovative practices which may inspire others.

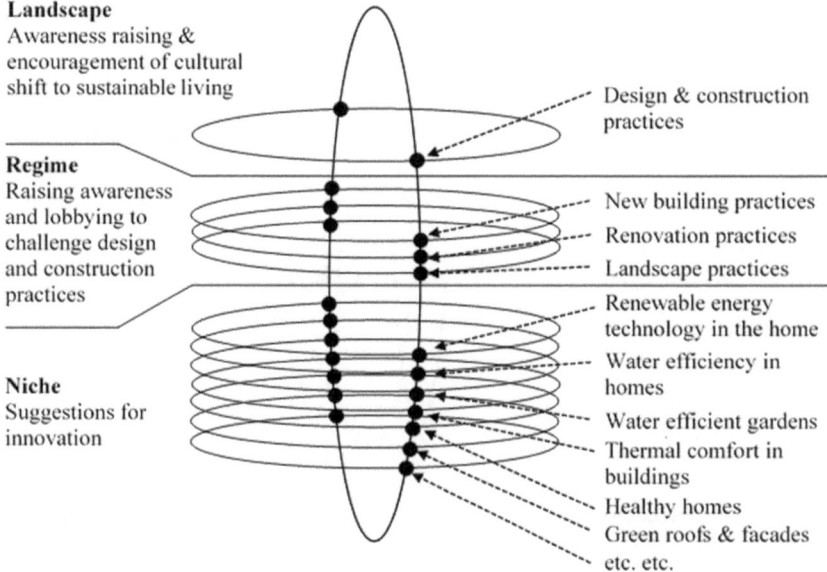

Fig. 6.2 Intersecting regimes and practices: Speed date a sustainability expert, Marrickville, NSW. (Source: Hillier 2021)

MAKING CONNECTIONS: PUBLIC EDUCATION AND AWARENESS OF INNOVATIONS

In an effort to limit the vulnerability of their local municipality and enhance its contribution to climate change, the former Marrickville Council (since amalgamated into the Inner West Council in NSW) has forged stronger connections with state and Federal governments, private businesses and the community in an attempt to help individuals and other stakeholders operate more sustainably. Bringing like-minded individuals and community groups into connection with each other to foster behaviour change and innovative practices has been a particular focus of the Council. One such initiative is their *Speed Date a Sustainability Expert* (SDSE) programme for local residents, undertaken in conjunction with Renew, the re-branded not-for-profit Alternative Technology Association (ATA). Operated successfully for a number of years across Australia, the SDSE programme connects individual residents with a broad range of sustainability experts for a 20-minute "date" where individualised advice can

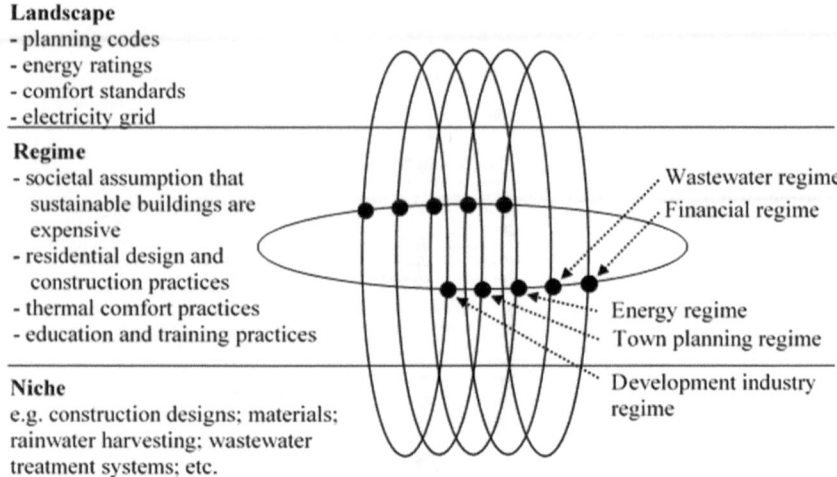

Landscape
- planning codes
- energy ratings
- comfort standards
- electricity grid

Regime
- societal assumption that
  sustainable buildings are
  expensive
- residential design and
  construction practices
- thermal comfort practices
- education and training practices

Niche
e.g. construction designs; materials;
rainwater harvesting; wastewater
treatment systems; etc.

Wastewater regime
Financial regime
Energy regime
Town planning regime
Development industry
regime

Fig. 6.3 Intersecting regimes and practices: Griff Morris—Solar Dwellings.
(Source: Hillier 2021)

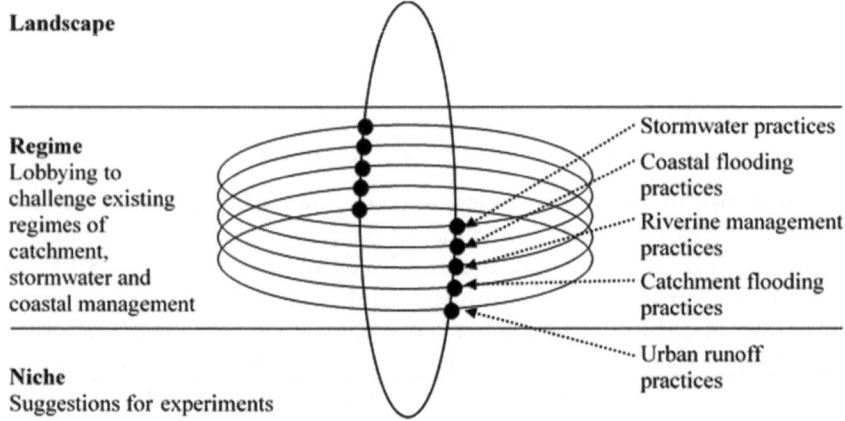

Landscape

Regime
Lobbying to
challenge existing
regimes of
catchment,
stormwater and
coastal management

Niche
Suggestions for experiments

Stormwater practices
Coastal flooding
practices
Riverine management
practices
Catchment flooding
practices
Urban runoff
practices

Fig. 6.4 Intersecting regimes and practices: Elwood floods action group.
(Source: Hillier 2021)

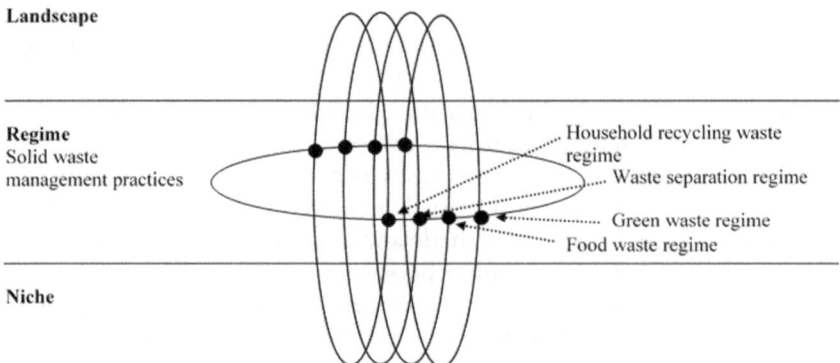

Landscape

Regime
Solid waste
management practices

.. Household recycling waste
regime
.... Waste separation regime

Green waste regime
Food waste regime

Niche

Fig. 6.5 Intersecting regimes and practices: Redland City Council—RedWaste. (Source: Hillier 2021)

be given on plans, sketches and ideas of renovation projects or new home designs.

The partnership between Renew/ATA and Marrickville came to be after the Council's Community Sustainability team recognised a demand from local residents for specific advice and information on sustainable solutions for renovation projects and new-build homes. In recent years, the Council's approach had shifted from being simply a regulator to pro-actively assisting residents in making more informed choices about their own sustainable changes. Recognising the limited knowledge Council could offer residents, the SDSE programme facilitated the marriage of local residents seeking specific advice for their projects, with experts who could guide local residents in designing, constructing and managing their own sustainable homes. The programme actively encourages new ways for Council, residents and experts to work together. Experts at Marrickville's 2016 SDSE event included specialists in permaculture and garden design, sustainable architecture and design, water collection, energy efficiency and passive solar design, off-grid solar, solar photovoltaics, energy storage, sustainable material selection, thermal performance, green roofs, walls and façades and so on.

By fostering new relations and facilitating the sharing of knowledge, Marrickville and Renew are creating new horizontal connections for local residents, enabling and empowering them to make informed decisions and build more environmentally sustainable and cost-effective homes. Each individual "date" enables the formation of new connections between experts and residents, triggering opportunities for behaviour change, innovative design solutions and the breaking of existing unsustainable renovation and home building regimes. Innovations are able to be incorporated from the start of a project, thus reducing costs. Facilitating the creation of these new connections eases locals' attempts to live more sustainably, mitigates the impacts of their building projects on the environment and adapts to changing climatic and environmental conditions.

The SDSE initiative emphasises personal practices in the private space of the home. Through their choice of new building and/or renovation practices, their landscaping practices and their selection of materials and processes, local homeowners can become empowered. SDSE questions traditional powerful systems, such as mainstream electricity generation, and plants elements of change; not in a grand gesture or outspoken form, but in a quiet manner of thinking and acting differently. It makes connections between people and experts, and between people, new technologies and practices. Putting SDSE into the Connections diagram, Fig. 6.2 demonstrates that it engages "niches" in particular, relating, for example, to renewable energy technology in the home, water efficiency in homes and gardens and so on.

Design and construction practices are challenged, with respect to new building and renovation as well as to garden landscaping as awareness is raised of sustainable practices. Connecting organisations, such as the former Marrickville Council and Renew, play important roles in facilitating points of intersection between practices and regimes. At the landscape scale, awareness raising and practice transformation may lead to a cultural shift to sustainable living, and eventually, perhaps, to changes in laws and regulations with regard to construction and development. SDSE is a bottom-linked form of social innovation which is continuing steadily Australia-wide. To date, there have been more than 50 SDSE events, several of them virtual during the COVID-19 period.

Breaking Connections: Reforming Institutions in Greener Directions

Since the late 1970s Griff Morris has been a dedicated champion in the research and design of sustainable, passive solar homes in Victoria and subsequently in Western Australia. When Morris designed and constructed his first passive solar home in 1977, his vision was to create a company which would raise awareness and knowledge of sustainable living throughout the building industry and the community. In 1991 Morris founded Solar Dwellings, with the aim of designing and constructing tailor-made, passive solar homes that are universally accessible and financially attainable for the consumer. Solar Dwellings' sustainable designs focus on creating individualised homes that reduce energy and water costs, are low allergen, multi-functional and multi-generational: pushing beyond compliance with the standard 6-star rating system and offering homeowners a sustainable, yet affordable, option. With construction costs at approximately $1000 a square meter for a standard four-bedroom, two-bathroom home, and $1400 per square metre for a two-storey 8-star design, Solar Dwellings demonstrates that energy efficient, sustainable houses can realistically cost the same as their non-efficient, non-sustainable equivalents. Solar Dwellings disrupts the assumption throughout the building industry that passive solar designs are unaffordable for the consumer, and unprofitable for the builder.

Griff Morris' passion for actualising affordable, sustainable homes—affordable to purchase and to maintain—is reflected in Solar Dwellings' mission statement: 'Making a difference to the environment in everything we do'. The company is committed to:

- Living and working sustainably
- Making a minimal impact on the environment
- Raising the benchmark for sustainable residential design
- Inspiring innovation in the Western Australian building industry
- Providing the most up to date sustainability education for the building industry and the community (http://www.solardwellings.com.au/about-us/our-mission).

Griff Morris and Solar Dwellings embrace the philosophy of 'doing the planet a favour'.[10] All their homes have minimal impact on the environment: intelligent passive design means they are energy and water efficient, reducing energy costs and greenhouse gas emissions; they are also universally accessible; most are low allergen utilising low-toxicity building processes and products. Morris and Solar Dwellings have received over 50 awards since 1991, including GreenSmart smart housing, water efficiency and design concept awards as well as several Master Builders Association awards. Solar Dwellings aim to design 10-star energy-efficient homes and other buildings. They build "nothing under 8 stars". The main niche components of their passive solar design are:

- Siting and orientation of the home, with living areas and large windows facing north, minimal windows to the east and west, and with the long axis of the home within 15° east or west of north.
- Winter warmth and summer cooling achieved by positioning windows so the sun enters in winter but not during summer.
- Natural cooling achieved by window placement that allows cross ventilation.
- Stable internal temperatures, resulting from appropriate materials, such as double brick, concrete or stone, used in the right locations. Timber floors can absorb the sun's warmth in winter and release it back into the home in the evening, while external shading prevents summer sun entering.
- Insulation is a barrier to heat flow, while most home builders include ceiling insulation as standard, Solar Dwellings recommends ceiling, under-roof and some wall insulation to retain winter warmth and exclude summer heat.
- Intelligent landscaping can protect a home from summer heat and maximise access to winter sun.

'We can further increase sustainability of a home through additional "smart home" measures', Morris says.[11] These innovations or niches include building materials which have low embodied energy and low environmental impact, so that a home is environmentally responsible in construction as well as in operation. Niches include non-chemical termite prevention, rainwater harvesting and other water efficiency measures, low-allergen and non-toxic building processes and products, active solar design, including photovoltaic power generation and solar water heating, aerobic sewerage and waterless urinals, grey water recycling systems, universal accessibility.[12]

Griff Morris is also the inaugural board member of the Sustainable Energy Association. He is a member of the Housing Industry Association (HIA) and a chief judge of its Housing Awards and the GreenSmart Awards. He is a member of the Commonwealth Government Committee regarding the Technical Advisory Manual for the Department of Climate Change and Energy Efficiency and is also a regular speaker at many public forums in WA, including SDSE. Morris has taught the HIA GreenSmart

Professional Process for some 15 years, together with sustainable design at the University of Western Australia. He is also a member of two Disability Services Commission committees and an ambassador for the Disability Services Commission *Count Me In Project.*

Griff Morris, as a quiet activist or "creative disruptor", has advocated for industry change by participating in various committees and industry bodies as above. Acting as an infiltrator, Morris takes his commitment for energy-efficient, accessible sustainable housing to the core of the building industry by sitting on the Environmental Planning Committee for the HIA. As the "fox among the wolves", he acknowledges that advocating for change to industry practices from within is not without its challenges, stating 'it needs a lot of grease to actually get some of those gears shifted'.[13]

For Morris, being in the space to allow for conversations that can shift attitudes, break down assumptions and initiate innovative practices is critical to breaking the existing socio-technical regimes that frame passive solar design as unaffordable, inaccessible and unprofitable. As he notes, the key to change is finding champions within, who can 'find the weak spot and then drill in as deep as they can'. Being the voice on the Environmental Planning Committee to query existing practices allows Morris, as an "infiltrator", to disrupt the existing regimes and practices of the building industry and to institutionalise change across the industry.

Griff Morris believes in the need for culture change; to disrupt and break the practice regimes which have served to stabilise the planning and development industry and the ways in which people regard their homes. For instance, he suggests that there needs to be more consciousness about "needing" and using air conditioners. People automatically look to turn on an air conditioner when the weather gets warmer. If people can be informed, they might break the practice of unsustainable living and come to demand affordable, sustainable houses. Similarly, there is a need to disrupt local and state governments' and planning authorities' lack of commitment to sustainability. This could be achieved through more informed interpretation of design codes, or even a basic understanding of the difference between solar access and light access.

To this end, Morris has been pushing and pushing for years on the committee 'to actually have the local government have a coordinated training process for their building surveyors...'. In addition, he suggests that every new estate development, 'because they all have a display area, every single one of those display areas should, at the start of the development, have to show how that home is efficient and why it is efficient'.

Then you follow through: 'you incentivise it for the salesperson. In other words, if they sell a home over 8 stars they get a bonus'.[14] In other words, encourage and persuade salespeople to sell what is best and not what is easiest to consumers.

Morris has also "been pushing" building and construction practices: 'you've got to get through to the builders'. 'It needs an overall strategy where you target the champions of industry. The people who can make change. The people in local government who have shown a commitment to it. The people in industry, whether it has been in the planning industry, certain planning groups as well as the developers themselves. And then, you inform those people, but you make sure that your information is fantastic and at the same time you go out into the public'.

Griff Morris' niche practices and products are shown in Fig. 6.3 which illustrates the points of intersection between Morris' sustainable building practices and regimes relating to the town planning and development industries, energy and wastewater systems, as well as those financial regimes which support (or otherwise) homebuyers and renovators and the residential construction industry. Morris disrupts and breaks connections of assumptions and practices at regime scale and also aims at breaking connections at landscape scale, for instance through his work on the Australian Commonwealth technical advisory manual.

MAKING CONNECTIONS: DIRECT LOCAL ACTIVISM

Contrasting with the bottom-linked quiet activism of SDSE and Griff Morris above, the Elwood Floods Action Group (EFLAG) offers a bottom-up example of quiet, but direct activism. One of the impacts of climate change is the occurrence of more severe and more frequent storms which increase the risk of flooding throughout river catchments and coastal areas. This is particularly true of the Elster Creek catchment in the inner south-east of Melbourne, Victoria. EFLAG is a good example of a self-organised, self-funded community organisation which illustrates how 'impulsive collective action [can] constitute an expression of creative, dynamic and engaged citizenship'.[15] As demonstrated below, EFLAG's activism has had an impact, triggering change in governance practices and flood management of Elster Creek.

The suburb of Elwood has a population of about 15,500 with a high median personal income and education attainment and a strong sense of community. Elwood is located on the coast where Elster Creek runs into

Port Phillip Bay. The lower sections of the area were once swampland. Efforts at reclamation began in 1859, although the area was only fully developed since the 1920s, with canals constructed and reconstructed across the following decades to facilitate drainage. Constructed drainage systems associated with increasing urbanisation gradually built on natural watercourses, with water runoff patterns further disrupted by the hard surfaces (roads, driveways and so on) of urban development. Despite concerted engineering attempts, the area has remained vulnerable to inundation, with the flooding caused by severe weather events in 2011 leading to the establishment of EFLAG.

To the members of EFLAG, the 2011 extreme weather event demonstrated a lack of preparedness on the part of government and other traditional service providers, undermining their ability to respond to the area's risk factors in an effective manner. The group sought to forge new and useful connections with stakeholders in a way that challenged the traditional top-down pattern of engagement. The ultimate aim of EFLAG was to make Elwood more resilient to extreme weather events through the precipitation of changes in practices across regimes. In particular, EFLAG identified links between the previously unlinked practices of riverine management and catchment flooding which occurred as a result of urbanisation and increased urban runoff into the river catchment, with coastal flooding practices linked to king tides, and with stormwater management practices, including those for flash flooding.

An essential element of EFLAG has been community engagement, encouraged through the holding of public meetings, leveraged to achieve their goals. EFLAG have adopted a forthright advocacy style that seeks to directly confront issues, ensuring a presence at all relevant meetings and forming partnerships with other concerned actors. Responsibility for drainage management in the Elster Creek catchment is complex, spread across the Victorian state government, state-owned Melbourne Water, four local government municipalities and individual property-holders. EFLAG identified the lack of coordination between agencies as a distinct obstacle to flood risk management. They, therefore, sought to foster constructive dialogue between what a respondent termed "the many-headed hydra": members of the community and decision-makers, including Melbourne Water, the State Emergency Service, City of Port Phillip, the Victorian chapter of the River Basin Management Society and so on.

Contact with the Cooperative Research Centre for Water Sensitive Cities (CRCWSC), an Australian government, university and industry partnership, resulted in Elwood becoming a case study for several projects relating to flood risk management.[16]A key element of the projects was a community envisioning process, attended by several members of EFLAG who made substantial contributions. Approximately six months later, a Working Group of key agencies in the Elster Creek catchment was established with specific objectives to 'explore the potential of a collaborative catchment-wide approach to develop consistent policy across the catchment, share information and explore technical solutions with potential catchment-scale benefits'.[17]

A Memorandum of Understanding was signed (and has been renewed) by the CEOs of Melbourne Water and the four local Councils and the Elster Creek Catchment Forum established in 2017, which in turn created the 2018 Elster Creek Action Plan. The Action Plan focused on short-term actions, including development of a community engagement strategy, and fed into development of the *Elster Creek Catchment Flood Management Plan 2020–2024*.[18] The Flood Management Plan acknowledges the community's input and advocacy of a collaborative approach to integrated water management, adopted not only in the Plan, but also in Amendment VC154 to the State Planning Policy in 2018. In addition, the Plan specifies actions including development of a communications and engagement plan involving key stakeholders (such as the State Emergency Service) and "community champions" and advocates to the insurance industry regarding reduced premiums for flood-resilient properties (a major issue raised by EFLAG).

In this way, previously isolated actors were brought into association with each other in order to create more effective cross-jurisdiction, multi-tiered government approaches to flood risk management, whilst ensuring the voice of the community remained heard. Vertically connecting the Elwood community into the urban water management landscape, these new connections and collaborations have enabled institutional change, the uptake of innovative practices by local authorities and a community with greater awareness and capacity to adapt to increased flood risks in a period of climate change.

Amongst the changes to which EFLAG has contributed, a respondent highlighted increases in local spending on drainage, changes to stormwater management practices, the utilisation of rain gardens for flood mitigation, the implementation of a monitoring system in the Elwood Canal that allows early warnings to be sent when water reaches a specific level, as well as ongoing consultation on policy and practices with a number of government and private entities. EFLAG has developed a plan for maximising water retention in the catchment and recycling, which is now utilised by Melbourne Water and local authorities. They have also developed proposals for cleaning Elwood Canal and restoring Elster Creek to facilitate stormwater to flow more freely, for duplicating a key drain and constructing a retardation basin.

EFLAG's aim appears to have been achieved. The group has not updated its website recently and appears to have dissolved, in the manner, perhaps, of "creative dissolution".[19] Nevertheless, EFLAG does demonstrate that bottom-up socially innovative activism can have strong transformative potential, reconfiguring governance structures. There have been positive changes to vertical connections between regimes and the broader landscape that exist as a result of EFLAG's work. The group has also encouraged more productive horizontal relationships between the four proximate local Councils and especially with the community. It regards Melbourne Water now convening catchment authorities with a range of representatives to plan cross-jurisdiction catchments as a major success. 'If you know the organisation structure and develop relationships with people that's how you get stuff done'.[20]

Putting EFLAG into the Connections diagram, Fig. 6.4 shows that there have been changes to the broader socio-technical landscape of flood management practices which now consider inter-jurisdictional issues. EFLAG's suggestions, for lots to have individual water tanks and so on, were novel contributions or "niches" not included in the 2019 *Flood Management Plan*. However, significant and successful innovation has occurred at the regime level where points of intersection have been identified and connections made between several practices and regimes.

EFLAG 'helped to reframe flooding problems in ways that were most relevant for the local community, came up with many creative and strategic ideas for technical, design and policy adaptation and recognised their role in contributing to flood resilience'.[21]

Breaking Connections: Confronting and Changing Institutional Practices

Redland City Council covers an area of 537 square kilometres, including mainland suburbs and six island communities in south-east Queensland on Moreton Bay. As a coastal community, the local government area of

Redland is particularly vulnerable to the effects of climate change. Awareness of this has led the council to involve itself, since about 1998, in schemes such as the ICLEI's (International Council for Local Environmental Initiatives), Cities for Climate Change Program. Funding from the Australian Federal Local Adaptation Pathways Program (LAPP) allowed the completion of a climate risk assessment. From that adaptation plan, the Council's Climate Change Strategy—*Confronting our Climate Future*—was developed.[22] This strategy included a commitment to reduce the waste of recoverable resources. Under the slogan '75 by 50', Redland reset its corporate greenhouse emissions targets to 25% lower than 1998 emissions by 2020, 50% lower by 2030 and 75% lower by 2050. These targets translate to reducing emissions by an average of 5% per year, every year.

At about the same time, the Redlands 2030 Community Plan *Creating our Future* incorporated a keynote "Green Living" theme, whose ten goals included:

- Goal 1 A culture of sustainability. Redland's citizens, communities, business and government are world leaders in understanding and committing to positive action to protect the future of the planet.
- Goal 2 Behavioural change. Redland's people take personal responsibility for carefully selecting, buying, using and disposing of the materials and services which support a sustainable lifestyle.
- Goal 4 Clean land, water and air. Garbage, pollution, contaminated stormwater and greenhouse gas emissions are minimised, not dumped into the environment for others to clean up.
- Goal 9 Leading waste management practices. Governments, businesses and residents reduce waste disposed to landfill by generating less waste, reusing waste materials, recycling or disposing of waste in ways that unlock or recover energy.[23]

Concerned about issues of energy consumption and greenhouse gas emissions, Redland decided to break with traditional ways of working with regard to solid waste disposal. Reducing, reusing and recycling waste was regarded as an opportunity to divert resources from landfill. The Council established RedWaste as a new business unit responsible for delivery of solid waste management collection and disposal activities across the mainland and islands. RedWaste's key functions include collection of kerbside waste and co-mingled recyclables from approximately 55,000 properties;

optional kerbside green waste collection (mainland only); litter bin collection from streets, parks and reserves; provision of eight Waste Transfer Stations that receive other waste and resources generated within the Redland Council area; and Recycle World tip shop that sells reusable items.[24]

RedWaste operates as a type 2 commercial business unit under local government legislation and includes both waste operations and waste planning units. Full cost pricing applies to RedWaste services, aiming to provide a commercial financial return to Council. In addition, the establishment of RedWaste brought an immediate decrease in charges for domestic consumers. Redland Council's ambitious *Sustainable Resources from Waste Plan 2010–2020*[25] commenced well. The Council's 2013–2014 Annual Report highlighted:

- The 10-year public place recycling project was completed in May 2014, ahead of schedule
- An additional 37 recycling enclosures and 35 waste enclosures were installed in parks and streets
- A bulk recycling bin service started on North Stradbroke Island aimed at diverting 34 tonnes of recyclable materials from the Island's businesses, community hall and waste transfer station annually
- On 30 June 2014, the kerbside green organics collection service diverted 2094 tonnes of green waste from 7244 participating properties, sending material to a composting facility for processing into soil and mulch products
- RedWaste continued the RedSWAP waste school-based education project and the establishment of a "kids teaching kids" programme.

Focusing on domestic waste, the three target measures of Redland's *Waste Reduction and Recycling Plan 2015–2020* are increase diversion of green waste, minimise food waste and increase diversion of kerbside recyclables. There has been an enthusiastic take-up of green waste bins across the mainland, based on an average annual growth of 13.5% since 2011. Some 2094 tonnes of green waste were collected by the kerbside green waste service in 2013–2014. As the cost of processing green waste is cheaper than the cost of landfill, the Council receives a positive financial contribution, in addition to a recycling benefit.

Redland also piloted a project to reduce food waste in the general waste bins through community-based social marketing, the *Waste Not Want Not* campaign, launched in 2017. Since 2011, the percentage of recyclable material diverted from kerbside collection has increased from 21.93% to 26.9% in 2018–2019. The total domestic tonnage of waste sent to landfill per capita/year has reduced over the same timeframe from 437 tonnes per capita in 2010–2011 to 355 tonnes in 2018–2019. In addition, the Council's gas extraction system has been expanded to extract gas from landfill waste.[26] It should be noted, however, that green waste collection, timber collection and processing, and bulk waste collection were contracted out in 2018–2019. This may relate to the increasing and above-budget expenditure on those services in 2017–2018.

RedWaste is well on the way to meeting its performance targets. It has creatively disrupted former solid waste practices, particularly amongst local households and schools. Previous kerbside practices of throwing all comingled waste into one bin have been replaced, in many instances, by waste separation regimes including green waste regimes and food waste regimes made possible by the broader socio-technical landscape of RedWaste's strategic and operational framework which ranges from the National Waste Policy to Redland Council's Waste Plan and RedWaste's own Performance Plan (see Fig. 6.5. Note: diagram omits trade waste, construction and demolition waste, regulated waste, disaster waste, dumping and littering).

RedWaste is a top-down example of social innovation. A Redland City Council respondent believes that the success 'comes down to timing and budget'.[27] Being part of ICLEI's Cities for Climate Change Program and obtaining funds from the Australian government's Local Adaptation Pathways Program enabled a strong basis to take ideas further. But as the respondent comments: 'there has been a number of drivers but now it is the right timing, and the right climate, and the right sort of appetite to do something'. What the respondent did not emphasise, however, was the important role they played in persuading local government officers and elected members to create the new Business Unit in such a way.

Our story only covers Redland Council's RedWaste practices concerned with domestic or municipal solid waste. We recognise, as in the diagram, that domestic waste is only part of RedWaste's core business. For information on practices dealing with commercial and industrial waste or trade waste, construction and demolition waste, regulated waste, disaster waste and dumping and littering, please see the *Waste Reduction and Recycling Plan 2015–2020*.[28] The case of RedWaste could have deep implications for municipalities. The opening up of institutional spaces for social innovation in this manner demonstrates a capacity to transform relations and practices at the regime scale in an economic governance climate which generally demands higher productivity, more performance indicators and a requirement to do more with less.

"CRITIQUE, SUBVERT, REWORK"

Making and breaking connections represent provocative challenges to the status quo. They can be creative disruptors, such as reading the landscape of legislation in new ways (in this chapter, for example, Griff Morris using the Australian Nationwide House Energy Rating Scheme as a performance tool rather than a compliance system). They can bring missing actors into conversations (e.g. Morris getting onto the Housing Industry Association environmental committee to 'try to change them from the inside'). All work together differently—from regional alliances of local government authorities to departments within an authority, such as Marrickville in the Inner West Council, New South Wales (NSW).

The four cases in this chapter demonstrate quiet activism on the edge of existing paradigms through the redirection of individual and collective practices in innovative ways. Social innovation has become an increasingly popular, yet ambiguous concept. We engage social innovation as both process and practice to meet the needs of local communities.[29] Of particular note is Redland City Council's development of the socially innovative RedWaste services in the face of sustainability and fiscal challenges. RedWaste offers an example of how socially innovative enterprises can be developed within local municipalities to practice differently, with potentially strong implications for urban governance.

RedWaste fits Bragaglia's category of local social innovation decided "from above"[30] at City Council level. Most social innovation initiatives, in contrast, appear to be "bottom-up" practices (exemplified by EFLAG), which are often regarded as superior to "top-down" state-administered activities because of their potential for greater responsiveness, cultural sensitivity and local empowerment. But successful innovation can rarely be categorised as either top-down or bottom-up, but as both shaping and shaped by dynamic cooperation across scales.[31] Griff Morris' activism ranges from local to Federal in Australia, while SDSE events are coordinated across the country by a national organisation in partnership, predominantly with local municipalities, sustainability experts and architects and occasionally with universities, NGOs and private organisations. These initiatives could be more properly regarded as "bottom-linked" developed interactively between individuals and collectives.[32]

As the cases highlight points of connection may not always perform leverage. Despite the best-efforts success is not assured or guaranteed. Some innovations and practices will remain relatively localised; others may be adopted elsewhere. Some may become institutionalised in state agencies and policies and/or in corporate practices and strategies. Others may dissolve when they have achieved their aims or disappear through lack of resources or commitment. Changing economic, social and/or political circumstances may negatively affect practices, whilst there may be resistance to innovation from those with vested interests in old ways of thinking and doing. Following Frank Moulaert and Diana MacCallum:

> We recognise that some may think that these initiatives, and our faith in them, are naïve; that local actions are impotent against the tsunami of neoliberal capitalism …; that their impact is too scattered and therefore missing the collective punch to make things happen; that their focus is too micropolitical and can only affect policy within existing delivery systems.[33]

But responses to climate change cannot be put on hold, irrespective of the enormity of other global problems. We need to do what we can. Finding and acting on critical points of connection, as leverage points, may convert barriers to change to inspirations for change. Quiet or noisy, individual or collective, indirect or direct, every action counts. We turn now to our final chapter which focuses on the need for and role of quiet activism at the local scale in a climate of crisis and change.

NOTES

1. Hall, S. (2020). The personal is political: feminist geographies of in/austerity. *Geoforum*, 110, 242–251.
2. Ibid., p. 118.
3. Ibid. p. 159.
4. Scott, K., Bakker, C. & Quist, J. (2012) Designing change by living change. *Design Studies*, 33, p. 283.
5. Rip, A. & Kemp, R. (1998). Technological change. In Rayner, S. & Malone, E. (Eds.), *Human Choice and Climate Change: Resources and Technology*, Vol. 2. Columbus, OH: Battelle Press.
6. Wetherell, M., McConville, A., & McCreanor, T. (2020). Defrosting the freezer and other acts of quiet resistance: Affective practice theory, every-

day activism and affective dilemmas. *Qualitative Research in Psychology*, 17(1), 13–35.

7. Shove, E. & Pantzar, M. (2005). Consumers, producers and practices. *Journal of Consumer Culture*, 5, 43–64.

8. Although leverage points are frequently associated with systems theories, see, for example, Alter (2013) and Scott et al. (2012) for discussion of links between systems and practices. Alter, S. (2013) Is Work System Theory a Practical Theory of Practice? Systems, Signs & Actions, 7(1), pp. 22–48; Scott, K., Bakker, C. & Quist, J. (2012) Designing change by living change. Design Studies, 33, pp. 279–297.

9. See Pottinger, L. (2017). Planting the seeds of a quiet activism. *Area*, 49(2), 215–222.

10. Solar Dwellings, accessed on http://www.solardwellings.com.au/about-us/our-philosophy

11. Excerpts from interviews undertaken as part of the Australian Research Council (ARC) DP150100299 project

12. For more details see http://www.solardwellings.com.au/about-us/media-releases

13. Excerpts from three interviews undertaken as part of the Australian Research Council (ARC) DP150100299 project transcripts.

14. Excerpts from interviews undertaken as part of the Australian Research Council (ARC) DP150100299 project transcripts.

15. Hasanov, M., Sudama, C. & Horlings, L. (2019). Exploring the role of community self-organisation in the creation and creative dissolution of a community food initiative. *Sustainability*, 11(11) [3170]. p. 2.

16. Rogers, B. & Gunn, A. (2015). *Towards a water sensitive Elwood: A community vision and transition pathways.* Clayton, Victoria: CRC for Water Sensitive Cities, Monash University; Rogers, B. et al. (2020). An interdisciplinary and catchment approach to enhancing urban flood resilience: a Melbourne case. *Philosophical Transactions of the Royal Society A* 378. https://doi.org/10.1098/rsta.2019.0201

17. Ibid. p. 15.

18. Melbourne Water, City of Kingston, City of Glen Eira, City of Port Phillip & Bayside City Council (2019). Elster Creek Catchment Flood Management Plan 2019–2024. Melbourne: Melbourne Water. Available at https://www.melbournewater.com.au/building-and-works/all-projects/elster-creek-catchment-flood-management-plan. Accessed Oct 22 2020.

19. See Hasanov, M., Sudama, C. & Horlings, L. (2019). Exploring the role of community self-organisation in the creation and creative dissolution of a community food initiative. *Sustainability*, 11(11) [3170]. https://doi.org/10.3390/su11113170, p. 2.

20. Excerpt from interviews undertaken as part of the Australian Research Council (ARC) DP150100299 project transcripts.

21. Rogers, B. et al. (2020). An interdisciplinary and catchment approach to enhancing urban flood resilience: a Melbourne case. *Philosophical Transactions of the Royal Society A* 378. https://doi.org/10.1098/rsta.2019.0201, p. 20.

22. Redland City Council (2011). Confronting our Climate Future. Available at https://www.redland.qld.gov.au/downloads/file/1028/confronting_our_climate_future. Accessed Oct 22 2020.

23. Redland City Council (2010a). Redlands 2030 Community Plan, Creating our Future, accessed https://www.redland.qld.gov.au/info/20226/council_plans_and_financial_information/424/community_plan

24. Redland City Council (2015). Waste Reduction and Recycling Plan 2015–2020, accessed on https://www.redland.qld.gov.au/downloads/file/2164/waste_reduction_and_recycling_plan_2015-2020

25. Redland City Council (2010b). Sustainable Resources from Waste Plan 2010–2020, accessed https://yoursay.redland.qld.gov.au/waste-strategy/documents/26172/download

26. Redland City Council (2019). Annual Report 2018–2019, accessed on https://www.redland.qld.gov.au/info/20226/council_plans_and_financial_information/433/annual_report

27. Excerpts from interviews undertaken as part of the Australian Research Council (ARC) DP150100299 project.

28. Redland City Council (2015). Waste Reduction and Recycling Plan 2015–2020, accessed on https://www.redland.qld.gov.au/downloads/file/2164/waste_reduction_and_recycling_plan_2015-2020

29. See Moulaert, F., Martinelli, F., Swyngedouw, E. & Gonzalez, S. (2005). Towards alternative model(s) of local innovation. *Urban Studies*, 42(11), 1969–1990; Moulaert, F., MacCallum, D., Mehmood, A. & Hamdouch, A. (Eds.) (2013). *The International Handbook on Social Innovation*. Cheltenham: Edward Elgar; MacCallum, D., Moulaert, F., Hillier, J. & Vicari Haddock, S. (Eds.) (2009). *Social Innovation and Territorial Development*. Farnham: Ashgate.

30. Bragaglia, F. (2020). Social innovation as a 'magic concept' for policy-makers and its implications for urban governance. *Planning Theory*. https://doi.org/10.1177/1473095220934832

31. Moulaert, F. & MacCallum, D. (2019). *Advanced Introduction to Social Innovation*. Cheltenham: Edward Elgar.

32. Moulaert, F., MacCallum, D., van den Broeck, P. & Garcia, M. (2019). Bottom-linked governance and socially innovative political transformation. In Howaldt, J., Kaletka, C., Schröder, A. & Zirngiebl, M. (Eds.), *Atlas of Social Innovation: A world of new practices*. Vol. 2. Munich: Oekom-Verlag;

Garcia, M. & Pradel, M. (2020). Bottom-linked approach to social innovation governance. In van den Broeck, P., Mehmood, A., Paidakaki, A. & Parra, C. (Eds.), *Social Innovation as Political Transformation*. Cheltenham: Edward Elgar.

33. Moulaert, F. & MacCallum, D. (2019). *Advanced Introduction to Social Innovation*. Cheltenham: Edward Elgar.

CHAPTER 7

Quiet Activism *in* Climate Change

Abstract This chapter returns to questions around what it means to be an activist, or to "do activism" in the context of the climate emergency. Responding to climate change is a profound societal challenge and crisis that requires transformative change across society's institutions and structures. Our interest is at the local scale where we argue international, national and state policies are translated into practices, and community mobilization, preparation and responses to the anticipated impacts of the climate crisis predominantly occur. Modest acts of care, connection and creativity can be collectively and politically significant. Through purposeful, collective commitment to socially innovative practices, local communities are forging new political pathways in response to the climate crisis. This we argue is the power and potential of quiet activism *in* climate change.

Keywords Activism • Climate emergency • Crisis • Regenerative practices • Connection • Local community

© The Author(s), under exclusive license to Springer Nature Switzerland AG 2021
W. Steele et al., *Quiet Activism*,
https://doi.org/10.1007/978-3-030-78727-1_7

Why Not Loud?

Our focus in this book is on quiet activism—the transformative potential of socially innovative and engaged strategies and tactics and the ways in which these can help to create new social relations and different urban imaginaries to address the climate emergency at the local scale. Socially innovative practices in local communities and contexts offer an alternative approach to those dominated by existing mainstream institutions largely ill-equipped to deal with the challenges of climate change, not least because of the constraints that institutionalised risk management places on climate justice and action.

The climate emergency demands that we focus on the shared knowledge and new practices that lead towards the transformative change needed if a sustainable world is to be possible. This requires both the short- and long-term strategic action, resistance to the status quo and democratic coordination to take human settlement systems in a new and different direction. Communities, governments, the private sector, not-for-profit and higher education institutions such as universities need to orient their collective efforts towards addressing this greatest of societal challenges—climate change. This is a transformative imperative and a key identified challenge is the slow progress and piecemeal approach to reducing emissions.

> Whether indexed by the continual climb in extreme heat and humidity, the melting of Arctic ice, the eruption of unprecedented mega-fire events or the rapid degradation of ecosystems and disruption of human settlements, climate change is here...Much-needed transitions towards low carbon and well-adapted systems are emerging. But they are too piecemeal and slow relative to what is needed to avoid large scale cascading and compounding impacts to our planet.[1]

In the context of a climate emergency and "the need for speed", quiet activism can be framed as at best an oxymoron, and at worst simply inadequate for the critical task at hand. There is a tendency within climate activism to dismiss "quiet" activities at the local scale as merely a precursor to bigger, more effective (i.e. "louder") political action. Alternatively, everyday local activities can be seen as disempowering or conservative, serving to reinforce rather than resist or transform a neoliberal agenda which privileges the modern urban bourgeoisie and its emphasis on individualised roles and responsibility.

Screw Light Bulbs, for example, was written by Donna Green and Liz Minchin in what they describe as a moment of fatigue after reading one too many articles about the huge threats posed by climate change alongside articles about well-intentioned individual action including the need to 'change your light bulbs'. They called it their "light bulb moment". In the face of the climate emergency, they argue small individualised actions such as replacing light bulbs are just not enough to address the scale and magnitude of the climate crisis. This led them to their position of 'screw light bulbs, let's look at what we really need to do to get serious about tackling climate change'.[2]

A similar argument has been made about household recycling when the recycle bins get taken to the one landfill, or the worth of home-scale vegetable gardens that simply supplement a meat-eating diet still largely reliant on mass produced food high in pesticides, a legacy of land clearing and carbon emissions. So, what then is the point if the unsustainable system remains the same? This is what some describe as the "dark side" of transformation which includes the risks associated with discourses and practices 'that construct transformation as apolitical, inevitable, or universally beneficial, which has the potential to produce significant material and discursive consequences'.[3]

In the Australian context Prime Minister Scott Morrison invoked the "quiet Australians" as a moniker to describe those who voted his Liberal/Nationals into government in the 2019 Federal election, including policies that sought to reverse or reduce national action on climate change.[4] As journalist Stan Grant observed, this referred to,

> Retirees, middle-class parents, and those dependent on the mining industry for their livelihoods all felt they were in the firing line. Christian leaders now say that religious freedom was a sleeper issue that turned votes in critical marginal seats. Throughout the world, long-silent voices are making themselves heard and it is shaking up politics as usual. People are saying they want to belong, and they want their leaders to put them first.[5]

The very nature of a transformative agenda is that it pushes community-members and decision-makers towards structural or root causes of issues such as power imbalances, questions of equity and injustice and worldviews. It also brings to the surface choices between often-normalised incremental change agendas, and more systemic transformative ones. This is the argument of the Extinction Rebellion global movement, which emphasizes that we are in the midst of an extinction of our own making and must act now to halt biodiversity loss and reduce greenhouse gas emissions to net zero by 2025. They call on the state to declare a climate and ecological emergency and to work with other institutions and community groups to communicate the urgency for change. 'Our main point here was to build a mass movement, to get mainstream people involved in the climate struggle in a way they never have before'.[6]

The prospect of intentional transformational change prompts reflection on the nature of climate justice and action. Naomi Klein in her book *On Fire* describes the rise of intentional social movements—"a people's emergency"—which is putting pressure on governments to change their policies to support action on climate change in key areas such as energy, transport, cities and agriculture through green new deals. She argues that the activism we are seeing today builds on a history of large-scale efforts driven predominantly by self-identified environmentalists and climate activists, that is rapidly changing to include everyday citizens as activists for change.

As Mark Pelling, Karen O'Brien and David Matyas write, this is the concept of transformation either 'forced by systems failure or chosen in anticipation of collapse and movement to a novel social-ecological systems state through a political decision-point, incremental adjustment or resistance to change in development pathways'.[7] They continue:

> Transformation describes depth of change, but not its origin, breadth or trajectory…This requires an understanding of the power relations that will direct, block or distort outcomes and may reduce predictability… What is the relationship between transformation, incremental adaptation, stability and resilience, and how might these processes interact? How and where might transformation emerge and spread? In what ways does transformation provoke changes in the approaches taken by researchers and practitioners?[8]

In the previous chapters we outlined the rich array of diverse and socially innovative climate action at the local scale. Social innovation is a form of empowerment that typically entails the creation of new organisational structures, service delivery modes, products and activities that meet social needs, or provide social benefits by grounding them in the relations and experiences of excluded groups and previously unmet needs. In Chap. 6—making and breaking vertical and horizontal connections—we focused on the creative practices of residential designer-builder Griff Morris, Marrickville Council and Renew (the Alternative Technology Association), the Elwood Floods Action Group (EFLAG) and RedWaste, a Council Business Unit engaged in recycling solid waste diverted from landfill. This built on Chaps. 2, 3, 4 and 5 which included exploring the socially innovative activist practices of Climate for Change, Green Cross Australia, Environment House and the *ReNew* Initiatives, Gecko, Karl Mallon, Climarte, One Planet, CANWin and Mossvale Community Garden, Solar $saver and Ecoburbia.

This final chapter focuses specifically on the nature and role of quiet activism *in* climate change. In doing so we return to questions around what it means to be an activist, or to "do activism" in the context of the climate emergency. This involves the need for significant changes to practices that are considered "normal" including what constitutes effective and transformative climate activism. At the local scale, this is often a focus on energy and consumption practices—making the choice to consume less, or consume differently, in recognition that we may not be able to rely on such a steady supply of energy, water, goods and services in the future. It also includes production and trade practices, when people develop local networks for producing, preparing and exchanging food and/or other goods and services.

Whether framed as climate art, sustainable household or community gardening practices, participatory street planting, or community economy schemes to get solar panels on roofs or to start-up climate-friendly businesses, these local-scale practices subtly and creatively subvert the boundaries between climate action and adaptation. Focusing on the intimate, the experiential and the everyday, quieter acts of climate action can show how different communities build practices that are more flexible and open-ended and which can communicate with each other in novel ways.[9] In each of the cases highlighted in the previous chapters, the emphasis was on identifying how (in what ways, and to what effect) quiet activist practices are being developed and defined by a shared interest and ethic of care to taking action to address the climate emergency at the local scale.

Activism *in* Climate Change

Our aim in this book is to reinforce the need for, and importance of, a depth and diversity of different types of climate *activism/s* that work better together to fundamentally shift the climate status quo. In doing so we do not seek to reinforce an "us" versus "them" mentality between different types of activism. As this does more to divide communities of practice, than create the necessary conditions for transformative change to thrive. In order to address the crisis of climate change we need fundamental *societal* change. Dealing with the loss of Aboriginal sacred sites and settled coastlines, desolation of agricultural land, severe disruptions to industry, climate refugees, fossil fuel dependence and unprecedented threats to biodiversity are all inter-related and will be catastrophic to both people and planet.

Climate change is not—as is often expressed at the scale of national government—peripheral and at arms' length. A crisis that is still "out there" in the distant future.[10] Scientist Mark Maslin has described five key (and corrupt) pillars of climate denial. *Science Denial,* he argues, sees climate change as a natural process that has always occurred, and that contrary to what the Intergovernmental Panel on Climate Change report claims, the science on climate change is unsettled and reliant on models and evidence that are inaccurate, distorted or misrepresented. *Economic Denial* posits that even if climate change is occurring, it is too expensive to address without destabilising society to an unacceptable degree. The third form of denial outlined by Maslin is *Humanitarian* which focuses on the potential benefits to be gained from a climate changed globe such as warmer weather and more productive farming lands in some areas. *Political Denial* highlights the dangers of taking action if other countries are not, and to focus on national rather than global responsibilities. Finally, *Crisis denial* addresses the need to maintain the status quo—or life as we know it—and not rush into changes such as a shift away from fossil fuels to renewables, given uncertainty about the outcome as a perverse misuse of the precautionary principle.[11]

The pervasive impact of these pillars of denial is reflected in governance and policy that prioritises "business as usual". Climate change is just another wicked planning issue, high in rhetoric but producing little in the way of substantive shifts to policy or practice. These targets are soft and fuzzy including resilience and adaptive capacity, increasing education and awareness, and taking steps to raise capacity for climate change planning through risk management strategies. Climate change is certainly part of "the mix" of everyday activity but remains outside the sphere of disruptive shifts to the cultural status quo needed for meaningful—that is transformative—societal change.

Activism in all its forms by contrast operates *in* climate change and in doing so emphasises the immediacy and lived dimensions of climate change now and into the future. This understanding of climate change is linked with collective mobilisation and critical praxis and draws attention to the need for a profound change in the way society functions in order to address the crisis conditions of the climate emergency experienced at multiple scales. A sense of crisis and urgency in the "here and now" is a shared feature of the climate activist community, and recognition that addressing climate change requires fundamentally different ways of living and working within planetary thresholds if a sustainable future is still possible.[12]

Anthropologist Deborah Bird Rose described activist engagement with the "here and now" as both an ethical encounter and entanglement; as well as a moral engagement of the past in the present without retreating to the future in which 'current contradictions and suffering will all be left behind justified by references to the future'.[13] For Rose, the pathway towards reconciliation is an open journey—social, spatial and temporal—and it is still possible to reshape relations between people, place and environments. This is linked to climate change as part of the larger pathway towards human–nature reconciliation, and an invitation to build together the types of knowledge and action, as well as practices and connections that can lead to transformative societal change. The challenge is to do this without defaulting to a future where some of us can seek to be washed clean or as Rose evocatively describes 'to remake ourselves, and all who are enmeshed in our damage, as trackless ghosts'.[14]

Climate activism can and must take many forms to be effective, and the role of large protests that target the national and international scale is crucial. Within this broader climate activism milieu and people power is what we see as the vital concomitant role of "quiet" activism. In this book we describe this quiet form of climate activism as the co-produced, small-scale and socially innovative activities and practices that demonstrate disruption, ingenuity, creativity and political craft-building at the level of local community. This is the prefigurative politics and regenerative practices that seek to be both innovative in creating solutions to otherwise intractable problems at the local scale; but also, more inclusive and empowering than traditional approaches enabling new actors to contribute to localised solutions to the climate emergency through everyday activities and practices.

THE NATURE OF QUIET ACTIVISM/S

In the context of addressing the climate emergency, building climate-just city and regions and affecting transformative societal change, there is no one "right" activist model or pathway. As Sarah Maddison and Sean Scalmer highlight in their book *Activist Wisdom: Practical Knowledge and Creative Tension in Social Movements,* activists have worked to change society in a range of different ways for the greater good, 'for justice, equality, recognition and democracy'.[15] Many of these gains remain ongoing, elusive and/or fragile, particularly those related to First Nations reconciliation and sovereignty, the transition to renewable energy and the logging of old growth forests which all form part of addressing the climate emergency.

Social justice movements—even those that are most successful and visible—do not "just happen". They are an assemblage of smaller scale practice-based activities that together combine to make both the sum and the parts equally important. These activities are many and varied but include writing letters, attending meetings, liaising with the media, lobbying with politicians, negotiating, planning, strategizing, arguing, organizing, debating, persuading and so on. This is a certain type of activist wisdom—"practical knowledge"—which seeks to create, intervene, challenge and disrupt in order to change the status quo. This kind of knowledge is not abstract but embedded in stories and practices. This is both a great strength for activists (e.g. tactile, situated, grounded knowledge and practices), as well as being an area of potential weakness (e.g. practices and knowledge that are opaque, non-transparent, difficult to communicate and replicate). For Maddison and Scalmer,

> Activists do not wonder "what is democracy"; they try to discover a way of working together and making decisions. They do not search for "an ethics of difference"; they attempt to find a means of listening to others and taking account of different needs. Their insights are practical...Stories poured forth.[16]

A number of tensions lie at the heart of these activist stories including but not limited to tensions around unity and difference; organisation and democracy; expressive and instrumental action; counter-publics and the mainstream; the local and global; redistribution and recognition and hope and despair. Common characteristics of activists are integrity, compassion, sincerity and a belief that change can happen. Applying the critical impact work of geographer Ruth Machen is instructive here for thinking about the practical knowledge of activism as a transformative, reciprocal agenda. She outlines strategies for creating critical impact including: building knowledge, networks and tools; listening, supporting, representing and mobilising marginalised voices; and inspiring critical skills and new forms of critical engagement. Such modes lead to:

- Challenging out-dated policy
- Empowering community resistances
- Platforming different voices
- Nurturing new critical publics[17]

Machen's work raises critical questions about the political and discursive work that is performed by activists and in particular 'the relationship between knowledge translation and hegemonic power, and the way that this relationship shapes particular forms of neoliberal climate governance'. This focus draws particular attention to the need to change the dominant discourses and delivery modes, and the need to embrace the boundary work and knowledge politics required to confront the climate emergency. However, as Machen cautions this can be also co-opted, redirected, silenced and/or sabotaged as, 'critical approaches impassioned by a desire for social change and seeking to challenge the status quo … often face a more challenging pathway to impact'.[18]

Collective everyday practices and the diversity of quiet activist orientations have often been described within the context of hobbies, crafts and gardening. These range from individualised, home-centred pursuits, to larger political collectives and social movements. Fiona Hackney for example describes the activist potential of amateur domestic crafts as historically conscious, marginalised spaces that promote agency and new imaginaries. These are the hidden, "quiet" zones which exist outside of masculine, capitalist culture, also critiqued for being underhand, subversive or "crafty". Hackney highlights the power of quiet craft as socially engaged practice emerging in contemporary activist movements such as guerrilla knitting, yarn bombing, 'stitch and bitch' and punk DIY. She argues this is the radical potential of everyday life which can serve to reconfigure the nature of work and the ways living sustainably can be imagined and shared, defined and materialised in practice.[19]

Laura Pottinger takes this further to emphasize the embodied nature of everyday activist practices as acts of care and kindness to community (both human and non-human). Her interest is a 'dirt under the fingernails' kind of activism, which gains strength from its subversive nature, embodied quietness and commitment to practical action. She explores these knowledge practices by focusing on local-scale suburban seed savers: suburban gardeners who cultivate fruits and vegetables at home and then select and save seed to provide future generations of plants for themselves and others. This also includes the localized hubs and not-for-profit organisations who make the connection between individual seed savers and the community. These quiet acts of growing and sharing are also part of a broader movement to conserve biodiversity and challenge the corporate control of food systems.[20]

Recognition of the power of everyday activist practices as a post-capitalist politics has also been powerfully highlighted in the work of JK Gibson-Graham and other critical scholars and practitioners.[21] The belief that radical change is possible through everyday practices that prefigure 'the world we want to live in' is core to this transformative agenda. The *Community Economies Collective* adopts an anti-essentialist approach which recognizes the power and efficacy of things that might seem small and insignificant and remains open to the unexpected and the unknown. This approach affirms that lives unfold in a "pluriverse" rather than a universe where a range of solutions and strategies for change towards more sustainable futures exists, alongside the ongoing processes of learning through ethical relationships and transformation.[22]

From a different perspective and positionality the popularity of Tactical Urbanism highlights the capacity for citizen engagement in the creation and activation of the neighbourhood through small, local-scale tactical projects that rise and thrive in response to outdated policies and planning.[23] Across the world the power of spontaneous and creative bottom-up protest movements are challenging "business as usual" and highlighting how ordinary citizens have started to take back power and create change.[24] In *An Urban Politics of Climate Change*, Harriet Bulkeley, Vanessa Castan Broto and Gareth Edwards argue that central to the response to climate change are new modes of experimentation and the multiple sites and spaces that are emerging "off-plan".[25]

Quiet activism as socially engaged and innovative practice at the local scale provides a lens through which to view the radical potential of everyday life. Questions of agency, connectivity (both social and familial), creativity, care and community shape a space and a praxis where 'sometimes surprising, collective identities, agencies, and capacities have developed'.[26] This is the power of small, purposeful everyday practices—a politics of making and doing. A form of place-based engagement that emphasises effort and energy to transform everyday practices, by working to reconfigure networks and power relations so that alternative values can be explored and shared. This, as Paul Chetterton and Jenny Pickerell describe in *Everyday Activism and the Transition to Post-capitalist Worlds,* is how 'everyday practices are used to build hoped-for futures in the present, and that this process is experimental, messy and contingent, and necessarily so'.[27]

THE POWER OF QUIET ACTIVIST PRACTICES

This book focuses on the collective action and grassroots social movements that are necessary to help support the transformative change needed to address the climate emergency at the local scale. The focus here is on "making and doing" as embodied creative practice and an activist politics in motion. Small repetitive, localised networks and actions provide a filter for change, connection and creative responses to unsustainable lifestyles and practices as a purpose-filled contribution to progressive goals. These are "quiet" practices that are modest and community-oriented, but no less potent and significant—indeed we argue essential—as part of the regenerative praxis needed for a transformative climate of change. As Chatterton and Pickerell continue,

> While traditional academic accounts of activism emphasise vocal, antagonistic and demonstrative forms of protest, geographers have begun to expand the category of activism to include modest, quotidian acts of kindness, connection and creativity.[28]

In the Australian cases we have highlighted and explored in this book, the focus is on building community-based knowledge practices through making and doing, care and co-production. This involved creating, trialling, modelling and sharing diverse practices and politics of solidarity around climate justice and change at the local scale rather than an emphasis on "getting their voices heard". There was no one, singular model but rather a continuum of quiet activism/s that shift depending on the context and community and that span from reformist to radical and revolutionary, but in all cases weaving together threads of transformative change that emphasize justice, care, social innovation and inclusion to address the climate emergency at the local scale. This is place-based and knowledge-based practice that is not abstract or disconnected but engrained and embedded—grounded—in localised life and community.

To explore climate action at the local scale, our research focuses specifically on the relationships between social innovation and local climate action. We are particularly interested in local-scale climate initiatives and activities that seek to (1) meet genuine needs; (2) engage and empower the local community in the preparation and delivery of strategy and initiatives; (3) hold the potential for the transformation of social relations; and (4) prioritize those most vulnerable in the community (both human and non-human) (see Fig. 7.1). This is how "just" outcomes are framed by activists at the local scale in ways that are both socially innovative and inclusive.

A strong emphasis in our research project and this resulting book has been on the nexus between social innovation and everyday climate activist practices with a particular focus on the Australian suburban landscape. Unlike feminist interests in private, individualised activism (i.e. at work in the home), our attention in this book focuses on socially innovative activities undertaken in the public realm; enacted at the local scale; working in, across and between different parts of the community; seeking to create intentional change in a climate emergency by creating new imaginaries at

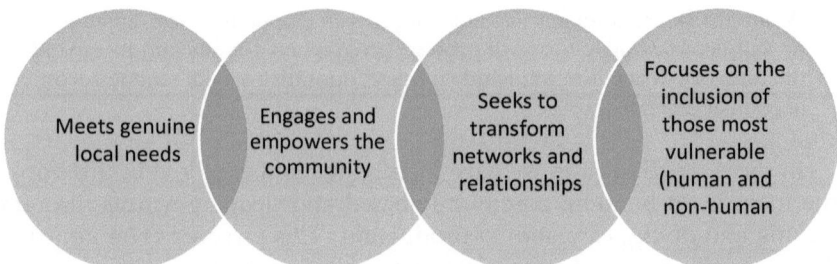

Fig. 7.1 Quiet activism: Key criteria of local scale climate practices. (Source: Authors)

the local scale; and purposively extending the realm of the political beyond cognitive and verbal into doing and making (see Fig. 7.2).

Collectively these quiet activist practices are underpinned by practices of care, cultivation and modesty at the local scale. The majority of our participants who self-selected for the research project were either white and/or middle class and this was a key recognised limitation. Further research is needed that explicitly and ethically addresses the urban edge and periphery where marginalisation and disenfranchisement through poverty, violence and racial, sexual or other forms of discrimination make social innovation far more difficult—and more critical. Following Simone de Beauvoir, 'it is in the knowledge of the genuine conditions of our lives that we must draw our strength to live and our reasons for acting'.

Five key themes emerged from the research that together underpin our understanding of "quiet activism": (1) building and bridging the knowledge base; (2) bringing missing actors to the table; (3) walking together with care; (4) realising transformative potential, and (5) making and breaking connections. Based on this, new practices can be identified that might be put to work in shaping climate action. These are performative, intentional and political, and emphasize different types of tactile, temporal and scalar activist practices.

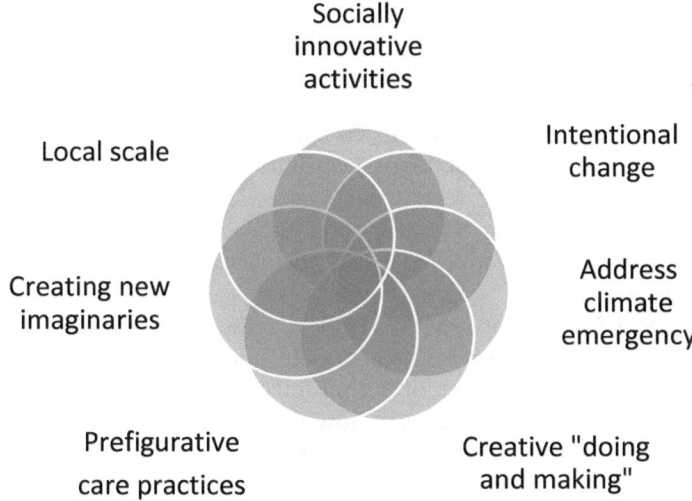

Fig. 7.2 The practices of quiet activism. (Source: Authors)

In Chap. 2 we focused on the first of these themes "Building and Bridging the Knowledge Base" and the power of stories in the activist process. How, in what ways and at what scale we construct the stories and practices of transformation around climate change are critically important.

> These narratives have what some social theorists call "performative effects." In other words, our narratives help to bring into being the worlds they describe. We are aware that the stories we tell can sometimes make the things we're trying to change seem more powerful and can therefore close off possibilities for change and dampen transformative inspiration ... It is therefore crucial that we cultivate representations of the world that inspire, mobilize, and support change efforts even while recognizing very real challenges.[29]

Climate for Change, for example, frames their work as a democratic project in citizen education and participatory climate action where attendees are asked to reflect on their own beliefs and values through conversation and are provided with specific ideas and actions they can pursue after the event. The group encourages people to go beyond individualistic action to amplify their impact as a collective, such as by hosting a subsequent gathering. This is a focus demonstrated in *Green Cross Australia* where "Think + Act + Share = Change" is central to their work which recognises that governments alone will not be able to deal with the scale of climate change–induced impacts, and there is an urgent need to build capacity, share knowledge and encourage practical action at the local scale. Whilst *Environment House*'s aim is to "walk the talk" *and* "talk the walk" on local-level climate protection, river and wetland care and action on climate change and sustainability issues. This is reflected in their constitution which commits them to trying to act on climate change at the local scale through community education and hands-on involvement.

Chapter 3, the second theme on "Bringing Missing Actors to the Table" emphasizes the activist practices that work to foster and enable community development practices and increase social inclusion. This in turn can build local capacity for action to climate impacts and increase involvement in diverse forms of community response-ability. The stories that map onto this theme illustrate how each dimension offers a different way to bring missing actors to the table. For instance, developing political agreement upon the broad principles of a local government strategy but putting the detail in supporting documents can bring politicians to the table to build effective leadership without engendering partisanship. Fostering community development practices (e.g. via multicultural networking days that increase social inclusion) can build capacity for adapting to climate impacts and increase resident involvement (e.g. neighbours checking on each other during extreme weather events) thus building response-ability. And new ways of bringing diverse actors together (e.g. trade unions, religious organisations and businesses)—such as practices of stewardship—can break through previous barriers to action.

As the profiled cases illustrate central to this is the idea of *enabling* innovation. Innovation This includes the key elements of enablement: the skills and practices of leadership; relationship building such as trust, commitment, open communication, collaboration and information exchange; as well as skills and competence development and decision-making support through problem identification, developing solutions, financial or human resourcing and delegating power to act. This approach to enablement is about response-ability across three key domains— behaviour change, institutional change and built form change in bringing missing actors to the table. While we sought to engage with a multiplicity of groups in the suburban heartlands of Australia, we did not explicitly seek out the experiences of the most marginalised and disadvantaged members of Australian society, people oftentimes dwelling in the interstices and margins. These groups include refugees, migrants, people with severe physical and mental disabilities, children, the elderly, homeless people, those with insecure tenure and the long-term unemployed, among others.

Nor did we work closely with First Nations people—Aboriginal and Torres Strait Islander Australians and the flourishing networks working at the frontline intersections of settler-colonialism and climate injustice for the benefit of all.[30] Despite our commitment to climate justice, critical reflexivity and ethics-based research practice, we recognise our unearned privilege as white settler researchers. It is critically important that people and communities, typically overlooked in mainstream climate change responses, are recognised. If we, as a society, are to truly build and bridge the knowledge base, bring missing actors to the table, walk together with care, foster enduring connections and networks of action and thus realise the transformative potential of social innovation, then important work remains to be done.

The need to "Walk Together with Care" was the third key quiet activism theme and the focus of Chap. 4 which highlights how many climate advocates working in government and community settings intuitively and creatively "work around" the politics of climate consensus and the politics of "fixing" climate action. There are strong parallels between everyday sustainability—the work that people do in their homes, schools, communities and workplaces to address carbon emissions in energy, food, fashion, household items with the quieter forms of activism that can be mobilized at the local scale. By focussing on the intimate, the experiential and the everyday, quieter acts of climate action can show how different communities build repertoires of adaptive practice that are more flexible and open-ended, and which can communicate with each other in novel and creative ways. Here, quiet activism, in its attentiveness to embodied and creative forms of ordinary doing and working, also inspires us to think about forms of quiet climate adaptation.

The common thread that runs through this chapter is that more carefull approaches of working-with and working-around climate-adaptive practices can be prioritised and privileged over "technical-risk" talk about climate mitigation and adaptation. These approaches build on what people care about and are actually doing through embodied, everyday practices. In putting the technical-risk narrative to the side, diverse ways of interweaving adaptation and mitigation with sustainable urban living are foregrounded. Each of the stories highlighted in this chapter—for example, *Cool Streets* in Blacktown City Council, NSW, or *CANWin* a non-partisan community group whose state focus is 'Working Together to Win/Climate Action Now'—profiles locally based community practices that take many forms including street tree planting to reduce urban heat islands, crowd-funding solar panels and sustainable businesses, and growing community connections through gardening and supporting local foods. The central element is the focus on the quiet and creative acts of care that can inspire not necessarily community consensus but new ways of walking together, differently.

An important corollary to the theme of 'walking together with care' is the activist agenda of 'realising transformative potential' which is a key theme outlined in Chap. 5 around socially innovative climate justice and action at the local scale. This is where international, national and state policies are translated into practices on the ground, and community mobilization, preparation and responses to the anticipated impacts of the climate crisis predominantly occur. In particular the focus is on the potential of local activists to expand their reach: to transform practices and social relations and empower actors more widely ("scaling out"), more formally ("scaling up") and/or more profoundly ("scaling deep") than their original scope and scale. As evidenced in the two cases explored, there are lots of different ways that these different kinds of scaling can happen.

Local initiatives are fundamentally shaped by their environmental, social and (especially) political contexts. Networking activities through which knowledge and practices can be exchanged and built are important for scaling out, and it was clear that the more diverse these activities are, the better. As such the processes of *scaling out* and *scaling up* are both vitally important, meaning the higher level institutionalisation of innovation by way of new or reinterpreted policy and public governance arrangements—that is, changes to the political context in ways that allow local experiments to sustain themselves and to flourish. This involves new "bottom linked" kinds of cooperation which are neither top-down nor bottom-up but emerge from the cooperation itself. This is *scaling deep*—having a profound effect on the lives and minds of the people who participate in these quiet climate activist initiatives.

Our final theme underpinning the practices of quiet activism at the local scale is "Making and breaking connections" as part of a prefigurative and regenerative politics for change. This chapter outlined innovative activist practices undertaken by local government authorities, non-government organisations and the private sector, and how they connect with other practices. In doing so we identified the critical points of connection and the potential footholds or leverage points for transformational change (i.e. praxis) in response to the climate crisis from designer-builders embedded in the Housing Industry Association; individuals at the former Marrickville Council and Renew (the Alternative Technology Association); local activists who helped to co-establish EFLAG, a local citizens' group in response to the 2011 Elwood floods, and local council officers at Redland City Council who persuaded colleagues and political members to form RedWaste, a Council Business Unit engaged in recycling solid waste diverted from landfill.

As these diverse stories demonstrate the connections that shape quiet activism at the local scale can be vertical and/or horizontal. Vertical connections can be distinguished by three "levels": macro-scale socio-technical landscapes, meso-scale socio-technical regimes and micro-scale niches. Horizontal connections include those that exist between local interest groups, or across regimes. The critical importance of making new connections for disseminating ideas and practices by breaking old connections works to disrupt unsustainable, outdated ways of thinking and acting. Making and breaking connections represent provocative challenges to the status quo. These activist practices can be creative disruptors, such as reading the landscape of legislation in new ways, or they can bring missing actors into conversations in order to engender transformative politicized change to address the climate emergency.

CLIMATE JUSTICE AND ACTION: HERE AND NOW

We are facing a climate emergency and responding to climate change is a whole of society crisis that requires transformative change across and between society's institutions, structures, space, places and experiences—not sometime in the future—but in the "here and now". We believe that quiet activism, undertaken at the local scale, is as important to the transformation of society as more visible and familiar forms of activism such as lobbying and mass protest. Transformative processes are complex, involving new relational practices between human, technological and other non-human actors (animals, plants, minerals, rivers, mountains, the weather, etc.). Participants in change-making practices gain collective resources (e.g. knowledge, social networks, infrastructures) which can be further mobilised to institutionalise, disseminate and/or deepen the impact of new practices. This is a critical aspect of empowerment for local communities in climate change, and both a precondition of socially driven change, and an important outcome of it.

However quiet activism is not a panacea. Like any other form of activism, collective action or large social movement, quiet activism holds the potential to be mobilised for regressive, conservative, perverse or maladaptive purposes. Without the support of an ethical framework and enabling relationships, many innovative ideas and practices will only be short lived or limited to small pockets of localised activity. However, this we argue is a feature of innovation and activism at all scales—and not just at the local scale of everyday lived experiences.

Throughout this book, we have drawn attention to an alternative approach to climate change responses—a local scale regenerative politics that seeks to re-shape the nature of relationships between people, places and environments. We have highlighted the potentialities of social innovation and suggested that socially innovative strategies can offer hope for collective action and may scaffold broader and/or deeper social change. But collective action needs to also be inclusive action. When we apply a critically reflexive lens to our own research approach, sampling and methods the theme of 'bringing missing actors to the table' highlights the importance of the academic activist to better attend to both the means *and* ends of the research life cycle, outcomes and process.

Through purposeful, collective commitment to socially innovative practices, local communities are forging new political pathways in response to the climate crisis. What is being described here is how working from everyday embodied practices with what people do and how they relate and make meaning; to a way of breaking down real and perceived barriers to addressing the climate emergency. Hope-full, care-full yet grounded practices amplify the capacity for transformative change at the local scale. Socially innovative practices can work by making a profound difference to the lives, minds, relations and feelings of those who participate. Everyday quiet activist practices can become embedded as a new normal at the local scale.

We need to do what we can—now. Societal responses to climate change cannot be put on hold, irrespective of the enormity of other global problems. Creatively participating in local climate action enables us to collectively re-imagine our experience of and responses to the climate emergency, laying the foundation for new possibilities of socially innovative initiatives and practices that lead to transformative change. This in turn underpins the prefigurative and regenerative practices needed to transform the status quo and address the roots of climate change as a shared planetary and societal trajectory. Modest everyday acts of care and connection can be collectively and politically significant. They are needed—and they matter.

For all those working in communities at the local scale to engage and inspire people to act "now" to address the climate emergency, this is the power and potential of quiet activism—and you are not alone.

Notes

1. Rickards, L. and Pietsch, T. (2020) Climate Change is the most important mission of universities in the 21st century, in *The Conversation,* January 5th, accessed on https://theconversation.com/climate-change-is-the-most-important-mission-for-universities-of-the-21st-century-139214
2. Green, D. and Minchin, L. (2010) *Screw Light Bulbs: Smarter ways to save Australians time and money,* Perth, UWA Press.

3. Blythe, J., Silver, J., Evans, L., Armitage, D., Bennett, N., Moore, M., Morrison, T., Brown, K. (2018) The darkside of transformation: Latent risks in contemporary sustainability discourse, *Antipode, 50*(5), 1206–1223.
4. Murphy, K. and Martin, S. (2019) Scott Morrison credits 'quiet Australians for miracle election, *The Guardian. 19 May, accessed on https://www.the-guardian.com/australia-news/2019/may/19/scott-morrison-credits-the-quiet-australians-for-miracle-election-victory*
5. Grant, S (2019) Scott Morrison Won Australia's Election Against All Odds. It Shouldn't Have Come as a Surprise, in *Foreign Policy*, May 21st, accessed on https://foreignpolicy.com/2019/05/21/scott-morrison-won-australias-election-against-all-odds-it-shouldnt-have-come-as-a-surprise-alp-shorten-folau-religion/
6. ABC (2019) *Extinction rebellion strains police resources*, accessed on https://www.abc.net.au/news/2019-10-13/extinction-rebellion-protests-disrupt-melburne-cost-thousands/11598086
7. Pelling, M., and O'Brien, K., Matyas, D. (2014) Adaptation and Transformation, *Climate Change*, Springer
8. Ibid.
9. Houston, D., MacCallum, D., Steele, W., Byrne, J. (2016) Climate Cosmopolitics and the Possibilities of Urban Planning. *Nature and Culture* 11(3), pp. 259–277.
10. Steele, W. (2020) *Planning Wild Cities: Human-Nature Relationships in the Urban Age*, New York/London, Routledge
11. Maslin, M. (2019) The five corrupt pillars of climate denial, in The Conversation, November 29, accessed on https://theconversation.com/the-five-corrupt-pillars-of-climate-change-denial-122893
12. See Steele, W. (2020).
13. Rose, D. B. (2004) *Reports from a Wild Country: Ethics for Decolonization*, University of New South Wales Press, Sydney.
14. Ibid., p. 178.
15. Maddison, S and Scalmer, S (2005) *Activist Wisdom: Practical knowledge and creative tension in social movements*, Sydney, NSW Press.
16. Ibid., p. 9.
17. Machen, R. (2018) Towards a critical politics of translation:(Re)producing hegemonic climate governance, *Environment and Planning E – Nature and Space*, 1(4), pp. 494–515.
18. Ibid.
19. Hackney, F. (2013) Quiet activism and the new amateur, in *the Journal of Design and Culture*, 5(2), pp. 169–184.
20. Pottinger, L. (2017) Planting the seeds of quiet activism, in *Area*, 49(2), pp. 215–222.

21. For further references see Leitner et al. 2008; Juris 2008; Routledge and Cumbers 2008; Doherty et al. 2007; Glassman 2002; Featherstone 2003 2008; Routledge 2003; Halfacree 2006; Chatterton 2005; Pickerill 2007; Routledge et al. 2007.
22. See Community Economies Research and Practice, accessed on https://www.communityeconomies.org/about/community-economies-research-and-practice
23. See for example Tactical Urbanist's Guide to getting it Done, accessed on http://tacticalurbanismguide.com/about/
24. Harden, J. (2013) *Quiet no more: New political activism in Canada and around the globe,* Vancouver, Lorimer.
25. Bulkeley, H., Castan Broto, V. and Edwards, G. (2015) *An Urban Politics of Climate Change: Experimentation and the Governing of socio-technical transitions,* London, Routledge.
26. Ibid.
27. Chatterton, P. and Pickerill, J. (2010) Everyday activism and transitions towards post-capitalist worlds, in *Transactions of the Institute of British Geographers,* 35, pp. 475–490.
28. Ibid.
29. See Community Economies Research and Practice, accessed on https://www.communityeconomies.org/about/community-economies-research-and-practice
30. Birch, T. (2017) *Climate Change, Recognition and Social Place-Making.* Sydney Review of Books. Accessed on https://sydneyreviewofbooks.com/essay/climate-change-recognition-and-caring-for-country/

BIBLIOGRAPHY

Abrahams, G., Johnson, B., & Gellatly, K. (2016). *Art+climate=change*. Melbourne University Press.

Agyeman, J., Schlosberg, D., Craven, L., & Matthews, C. (2016). Trends and directions in environmental justice: From inequity to everyday life, community, and just sustainabilities. *Annual Review of Environment and Resources, 41*, 321–340.

Alam, A., & Houston, D. (2020). Rethinking care as alternate infrastructure. *Cities, 100*, 1–10.

Ambrey, C., Byrne, J., Matthews, T., Davison, A., Portanger, C., & Lo, A. (2017). Cultivating climate justice: Green infrastructure and suburban disadvantage in Australia. *Applied Geography, 89*, 52–60.

Baker, S., & Mehmood, A. (2015). Social innovation and the governance of sustainable places. *Local Environment: The International Journal of Justice and Sustainability, 20*(3), 321–334.

Belenky, M., Clinchy, B., Goldberger, N., & Tarule, J. (1986). *Women's ways of knowing: The development of self, voice and mind*. Basic Books.

Birch, T. (2017). *Climate change, recognition and social place-making*. Sydney review of books. Accessed on https://sydneyreviewofbooks.com/essay/climate-change-recognition-and-caring-for-country/

Blythe, J., Silver, J., Evans, L., Armitage, D., Bennett, N., Moore, M., Morrison, T., & Brown, K. (2018). The darkside of transformation: Latent risks in contemporary sustainability discourse. *Antipode, 50*(5), 1206–1223.

Boggs, C. (1977). Revolutionary process, political strategy, and the dilemma of power. *Theory and Society, 4*(3), 359–393.

© The Author(s), under exclusive license to Springer Nature Switzerland AG 2021
W. Steele et al., *Quiet Activism*,
https://doi.org/10.1007/978-3-030-78727-1

Booth, K., & Kendal, D. (2020). Underinsurance as adaptation: Household agency in places of marketisation and financialization. *Environment and Planning A: Economy and Space, 52*(4), 728–746.

Bragaglia, F. (2020). Social innovation as a 'magic concept' for policy-makers and its implications for urban governance. *Planning Theory*. Accessed on https://journals.sagepub.com/doi/abs/10.1177/1473095220934832

Brooks, S. (2014). Enabling adaptation? Lessons from the new 'Green revolution' in Malawi and Kenya. *Climatic Change, 122*(1–2), 15–26.

Bulkeley, H., & Betsill, M. (2005). Rethinking sustainable cities: Multilevel governance and the 'urban' politics of climate change. *Environmental Politics, 14*(1), 42–63.

Bulkeley, H., Castan Broto, V., & Edwards, G. (2015). *An urban politics of climate change: Experimentation and the governing of socio-technical transitions.* Routledge.

Burch, S. (2010). Transforming barriers into enablers of action on climate change: Insights from three municipal case studies in British Columbia, Canada. *Global Environmental Change, 20*, 287–297.

Byrne, J., Gleeson, B., Howes, M., & Steele, W. (2009). Climate change and urban resilience: The limits of ecological modernization as an adaptive strategy. In S. Davoudi, J. Crawford, & A. Mehmood (Eds.), *Planning for climate change: Strategies for mitigation and adaptation for spatial planners* (pp. 136–154). Earthscan.

Chatterton, P., & Pickerill, J. (2010). Everyday activism and transitions towards post-capitalist worlds. *Transactions of the Institute of British Geographers, 35*, 475–490.

Clarke, K. (2016). Willful knitting? Contemporary Australian craftivism and feminist histories. *Continuum, 30*(3), 298–306.

Climate Council of Australia. (2020). *Summer of crisis report.* Accessed online: https://www.climatecouncil.org.au/wp-content/uploads/2020/03/Crisis-Summer-Report-200311.pdf

Conradson, D. (2003). Spaces of care in the city: The place of a community drop-in Centre. *Social & Cultural Geography, 4*(4), 507–525.

Deleuze, G., & Guattari, F. (1987). *A thousand plateaus: Capitalism and schizophrenia.* Athlone Press.

Dowler, L., Cuomo, D., Ranjbar, A., Laliberte, N. C., & J. (2019). Care. In *Keywords in radical geography: Antipode at 50 antipode foundation.* Wiley-Blackwell.

Dupuis, J., & Knoepfel, P. (2013). The adaptation policy paradox: The implementation deficit of policies framed as climate change adaptation. *Ecology and Society, 18*(4), 31.

Eisenmann, L. (2005). A time of quiet activism: Research practice, policy in American women's higher education 1945–1965. *History of Education Quarterly, 45*(1), 1–17.

Flannery, T. (2020). *The climate cure: Solving the climate emergency in the era of COVID-19.* The Text Publishing Company.

Frost, J. S., Currie, M. J., Northam, H. L., & Cruickshank, M. (2017). The experience of enablement within nurse practitioner care: A conceptual framework. *The Journal for Nurse Practitioners, 13*(5), 360–367.

Garcia, M., & Vicari Haddock, S. (2016). Special issue: Housing and community needs and social innovation responses in times of crisis. *Journal of Housing and the Built Environment, 31*(3), 393–407.

Geels, F. (2011). The multi-level perspective on sustainability transitions: Responses to eight criticisms. *Environmental Innovation and Societal Transitions, 1*, 24–40.

Geels, F. (2014). Regime resistance against low-carbon transitions: Introducing politics and Power into the multi-level perspective. *Theory, Culture and Society, 31*(5), 21–40.

Gibson-Graham, J. K. (2006). *A Postcapitalist politics.* University of Minnesota Press.

Göpfert, C., Wamsler, C., & Lang, W. (2019). Institutionalizing climate change mitigation and adaptation through city advisory committees: Lessons learned and policy futures, *City and Environment Interactions, 1*, 1–12.

Grant, S. (2019, May 21). Scott Morrison won Australia's election against all odds. It shouldn't have come as a surprise. *Foreign Policy.* Accessed on https://foreignpolicy.com/2019/05/21/scott-morrison-won-australias-election-against-all-odds-it-shouldnt-have-come-as-a-surprise-alp-shorten-folau-religion/

Green, D., & Minchin, L. (2010). *Screw light bulbs: Smarter ways to save Australians time and money.* UWA Press.

Hackney, F. (2013). Quiet activism and the new amateur. *The Journal of Design and Culture, 5*(2), 169–184.

Hall, S. (2020). The personal is political: Feminist geographies of in/austerity. *Geoforum, 110*, 242–251.

Hao, Z., Singh, V. P., & Hao, F. (2017). Compound extremes in Hydroclimatology: A review. *Water, 10*(6), 1–24.

Harden, J. (2013). *Quiet no more: New political activism in Canada and around the globe.* Vancouver.

Hargreaves, T., Longhurst, N., & Seyfang, G. (2013). Up, down, round and round: Connecting regimes and practices in innovation for sustainability. *Environment & Planning A, 45*, 402–420.

Hasanov, M., Sudama, C., & Horlings, L. (2019). Exploring the role of community self-organisation in the creation and creative dissolution of a community food initiative. *Sustainability, 11*(11), 3170.

Head, L. (2016). *Hope and grief in the anthropocene.* Routledge.

Head, L., & Gibson, C. (2012). Becoming differently modern: Geographic contributions to a generative climate politics. *Progress in Human Geography, 36*(6), 699–714.

Healey, P. (2008). *Urban complexity and spatial strategies: Towards a relational planning for our times.* Routledge.

Hornsey, M., & Fielding, K. (2020). Understanding (and reducing) inaction on climate change. *Social Issues and Policy Review, 14*(1), 3–35.

Hotker, M., Steele, W., & Wiesel, I. (2019). When gambling fails: Caring-with urban communities at the local scale. *Cities, 100.* Accessed on https://www.sciencedirect.com/science/article/pii/S0264275119314222

Houston, D., MacCallum, D., Steele, W., & Byrne, J. (2016). Climate cosmopolitics and the possibilities of urban planning. *Nature and Culture, 11*(3), 259–277.

Houston, D., & Vasudevan, P. (2018). In R. Holifield, J. Chakraborty, & G. Walker (Eds.), *The Routledge handbook of environmental justice* (pp. 241–251). Routledge.

Howaldt, J., & Schwarz, M. (2016). Verifying existing social theories in reference to social innovation and its relationship to social change. *SI-DRIVE deliverable* 1.3.

Howaldt, J., Kaletka, C., Schröder, A., & Zirngiebl, M. (2019). *Atlas of social innovation, 2nd volume - A world of new practices.* oekom Verlag GmbH.

Hudon, C., Tribble, D., Bravo, G., & Poitras, M.-E. (2011). Enablement in health care context: A concept analysis. *Journal of Evaluation in Clinical Practice, 17*, 143–149.

Hulme, M. (2018). Weather-worlds of the anthropocene and the end of climate. *Weber: The Contemporary West, 34*(1), 59–70.

Ingold, T. (2010). Footprints through the weather-world: Walking, breathing, knowing. *Journal of the Royal Anthropological Institute, 16*(1), 121–139.

Intergovernmental Panel on Climate Change (IPCC). (2013). *Climate change 2013: The physical science basis.* Accessed on https://www.ipcc.ch/report/ar5/wg1/

Ireland, P., & McKinnon, K. (2013). Strategic localism for an uncertain world: A postdevelopment approach to climate change adaptation. *Geoforum, 47*, 158–166.

Jones, J., Winch, S., Strube, P., Mitchell, M., & Henderson, A. (2016). Delivering compassionate care in intensive care units: Nurses' perceptions of enablers and barriers. *Journal of Advanced Nursing, 72*(12), 3137–3146.

Karoly, D. (2015, May 7). Climate science is looking to art to create change. *The conversation*. Accessed on https://theconversation.com/climate-science-is-looking-to-art-to-create-change-41185

Kemp, R., Loorbach, D., & Rotmans, J. (2007). Transition management as a model for managing processes of co-evolution towards sustainable development. *The International Journal of Sustainable Development & World Ecology, 14*(1), 78–91.

Kenis, A., & Lievens, M. (2017). Imagining the carbon neutral city: The (post) politics of time and space. *Environment and Planning A, 49*(8), 1762–1778.

Kopp, R. E., Hayhoe, K., Easterling, D. R., Hall, T., Horton, R., Kunkel, K.E., & LeGrande, A. N. (2017). Potential surprises — compound extremes and tipping elements. In D. J. Wuebbles, D. W. Fahey, K. A. Hibbard, D. J. Dokken, B. C. Stewart, & T. K. Maycock (Eds.), *Climate science special report: Fourth national climate assessment, Volume I* (pp. 411–429). U.S. Global Change Research Program.

Latour, B. (2010). An attempt at a compositionist manifesto. *New Literary History, 41*(3), 471–490.

Lawson, V. (2009). Instead of radical geography, how about caring geography? *Antipode, 41*(1), 210.

Lucas, C. H., & Davison, A. (2019). Not 'getting on the bandwagon': When climate change is a matter of unconcern. *Environment and Planning E: Nature and Space, 2*(1), 129–149.

MacCallum, D., Byrne, J., & Steele, W. (2014). Whither justice? An analysis of local climate change responses from South East Queensland, Australia. *Environment and Planning. C, Government & Policy, 32*(1), 70–92.

MacCallum, D., Moulaert, F., Hillier, J., & Vicari Haddock, S. (Eds.). (2009). *Social innovation and territorial development*. Ashgate.

Machen, R. (2018). Towards a critical politics of translation: (re)producing hegemonic climate governance. *Environment and Planning E – Nature and Space, 1*(4), 494–515.

Maslin, M (2019, November 29). The five corrupt pillars of climate denial. *The Conversation*. Accessed on https://theconversation.com/the-five-corrupt-pillars-of-climate-change-denial-122893

Matthews, T., Lo, A. Y., & Byrne, J. A. (2015). Reconceptualizing green infrastructure for climate change adaptation: Barriers to adoption and drivers for uptake by spatial planners. *Landscape and Urban Planning, 138*, 155–163.

Moore, M., Riddell, D., & Vocisano, D. (2015). Scaling our, scaling up, scaling deep. *Social Innovation, 58*, 67–84.

Moulaert, F., MacCallum, D., & Hillier, J. (2013). Social innovation: Intuition, precept, concept, theory and practice. In F. Moulaert et al. (Eds.), *International handbook of social innovation*. Elgar.

Moulaert, F., MacCallum, D., Mehmood, A., & Hamdouch, A. (2013). General introduction: The return of social innovation as a scientific concept and a social practice. In F. Moulaert et al. (Eds.), *International handbook of social innovation*. Elgar.

Moulaert, F., MacCallum, D., Mehmood, A., & Hamdouch, A. (Eds.). (2013). *International handbook of social innovation: Collective action, social learning and transdisciplinary research*. Elgar.

Moulaert, F., MacCallum, D., van den Broeck, P., & Garcia, M. (2019). Bottom-linked governance and socially innovative political transformation. In J. Howaldt, C. Kaletka, A. Schröder, & M. Zirngiebl (Eds.), *Atlas of social innovation: A world of new practices* (Vol. 2).

Moulaert, F., Martinelli, F., Swyngedouw, E., & Gonzalez, S. (2005). Towards alternative model(s) of local innovation. *Urban Studies, 42*(11), 1969–1990.

Mueller, M. (2017). *Being salmon, being human: Encountering the wild in us and us in the wild*. Chelsea Green Publishing.

Murphy, K., & Martin, S. (2019, May 19). Scott Morrison credits 'quiet Australians for miracle election. *The Guardian*. Accessed on https://www.theguardian.com/australia-news/2019/may/19/scott-morrison-credits-the-quiet-australians-for-miracle-election-victory

Neumeier, S. (2017). Social innovation in rural development: Identifying the key factors of success. *The Geographical Journal, 183*(1), 33–46.

Nursey-Bray, M., Palmer, R., Smith, T. F., & Rist, P. (2019). Old ways for new days: Australian indigenous peoples and climate change. *Local Environment, 24*(5), 473–486.

O'Malley, N. (2020, December 1). World awaits action by 'suicidal' Australia, says former climate chief. *The Age*. Accessed on https://www.theage.com.au/environment/climate-change/world-awaits-action-by-suicidal-australia-says-former-climate-chief-20201201-p56joj.html

O'Donnell, E. L., & Talbot-Jones, J. (2018). Creating legal rights for rivers: Lessons from Australia, New Zealand, and India. *Ecology and Society, 23*(1), 7.

Oekom-Verlag, Garcia, M., & Pradel, M. (2020). Bottom-linked approach to social innovation governance. In P. van den Broeck, A. Mehmood, A. Paidakaki, & C. Parra (Eds.), *Social innovation as political transformation*. Edward Elgar.

Parés, M., Ospina, S. M., & Subirats, J. (2017). *Social innovation and democratic leadership: Communities and social change from below*. Elgar.

Pasquini, L., Ziervogel, G., Cowling, R. M., & Shearing, C. (2015). What enables local governments to mainstream climate change adaptation? Lessons learned from two municipal case studies in the Western cape, South Africa. *Climate and Development, 7*(1), 60–70.

Pears, A. (2017, August 6). Poor households are locked out of green energy, unless governments help. *The Conversation*. Accessed on https://theconversation.com/poor-households-are-locked-out-of-green-energy-unless-governments-help-81987

Pelling, M., O'Brien, K., & Matyas, D. (2014). *Adaptation and transformation*. Springer.

Potter, E., Miller, F. P., Lovbrand, E., Houston, D., McLean, J., O'Gorman, E., Evers, C., & Ziervogel, G. (2020). A manifesto for shadow places: Re-imagining and co-producing connections for justice in an era of climate change. *Environment and Planning E: Nature and Space*. Accessed on https://doi.org/10.1177/2514848620977022.

Pottinger, L. (2017). Planting the seeds of quiet activism. *Area, 49*(2), 215–222.

Power, E. (2019). Assembling the capacity to care: Caring-with precarious housing. *Transactions of the Institute of British Geographers, 2019*, 1–15.

Puig, de la Bellacasa, M. (2017). *Matters of care: Speculative ethics in more than human worlds*. University of Minnesota Press.

Rein, M., & Schon, D. (1994). *Frame reflection: Toward the resolution of intractable policy controversies*. Basic Books.

Rickards, L., & Pietsch, T. (2020, January 5). Climate change is the most important mission of universities in the 21st century. *The Conversation*. Accessed on https://theconversation.com/climate-change-is-the-most-important-mission-for-universities-of-the-21st-century-139214

Rip, A., & Kemp, R. (1998). Technological change. In S. Rayner & E. Malone (Eds.), *Human choice and climate change: Resources and technology* (Vol. 2). Battelle Press.

Rogers, B., & Gunn, A. (2015). *Towards a water sensitive Elwood: A community vision and transition pathways*. CRC for Water Sensitive Cities, Monash University.

Rogers, B., et al. (2020). An interdisciplinary and catchment approach to enhancing urban flood resilience: A Melbourne case. *Philosophical Transactions of the Royal Society A 378*. Accessed on https://doi.org/10.1098/rsta.2019.0201.

Rose, D. B. (2004). *Reports from a wild country: Ethics for decolonization*. University of New South Wales Press.

Rose, D., & Robin, L. (2004). The ecological humanities: An invitation. *Australian Humanities Review*, 31–32. Accessed on http://australianhumanitiesreview.org/2004/04/01/the-ecological-humanities-in-action-an-invitation/

Sandercock, L., & Forsyth, A. (1992). A gender agenda: New directions for planning theory. *Journal of the American Planning Association, 58*(1), 49–59.

Scannell, L., & Gifford, R. (2013). Personally relevant climate change: The role of place attachment and local versus global message framing in engagement. *Environment and Behavior, 45*(1), 60–85.

Schlosberg, D., Collins, L. B., & Niemeyer, S. (2017). Adaptation policy and community discourse: Risk, vulnerability, and just transformation. *Environmental Politics, 26*(3), 413–437.

Schultz, J. (2011). Life in a time of disasters. In *Surviving* (Griffith review, 35) (pp. 7–11). Griffith University.

Scott, K., Bakker, C., & Quist, J. (2012). Designing change by living change. *Design Studies, 33,* 283.

Scott-Cato, M., & Hillier, J. (2010). How could we study climate-related social innovation? Applying Deleuzian philosophy to transition towns. *Environmental Politics, 19*(6), 869–887.

Seyfang, G., & Haxeltine, A. (2012). Growing grassroots innovations: Exploring the role of community-based initiatives in governing sustainable energy transitions. *Environment and Planning. C, Government & Policy, 30,* 381–400.

Shove, E. (2003). *Comfort, cleanliness and convenience: The social organization of normality.* Oxford.

Shove, E., & Pantzar, M. (2005). Consumers, producers and practices. *Journal of Consumer Culture, 5,* 43–64.

Simons, M. (2020). *Cry Me a River,* The quarterly essay, Issue, 77. Black Inc Publishing.

Slezak, M. (2017, March 22). Renewables roadshow: How the 'nonna effect' got Darebin's pensioners signing up to solar. *The Guardian.* https://www.the-guardian.com/environment/2017/mar/22/renewables-roadshow-how-the-nonna-effect-changed-darebins-approach-to-solar

Solnit, R. (2013). *The faraway nearby.* Viking.

Steele, S., MacCallum, D., Byrne, J., & Houston, D. (2012). Planning the climate-just city. *International Planning Studies, 17*(1), 67–83.

Steele, W. (2020). *Planning wild cities: Human-nature relationships in the urban age.* Routledge.

Steele, W., & Rickards, L. (2021). *The sustainable development goals and higher education: A transformative agenda?* Palgrave.

Steffen, W., Dean, A., & Rice, M. (2019). *Weather gone wild: Climate change fuelled extreme weather.* Climate Council of Australia. Accessed on https://www.climatecouncil.org.au/wp-content/uploads/2019/02/Climate-council-extreme-weather-report.pdf

Stengers, I. (2005). The cosmopolitical proposal. In B. Latour & P. Weibel (Eds.), *Making things public: Atmospheres of democracy* (p. 1002). MIT Press.

Stengers, I. (2010). *Cosmopolitics I, II.* Trans. Bononno, R. University of Minnesota Press.

Stone, C. D. (1972). Should trees have standing? Toward legal rights for natural objects. *Southern California Law Review, 45,* 450–501.

Tompkins, E. L., & Adger, W. N. (2005). Defining response capacity to enhance climate change policy. *Environmental Science and Policy, 8,* 562–671.

Tronto, J. (2017). There is an alternative: Homines curans and the limits of neo-liberalism. *International Journal of Care and Caring, 1*(1), 27–43.

Tronto, J. C. (1993). *Moral boundaries: A political argument for an ethic of care.* Psychology Press.

Tronto, J. C., & Fisher, B. (1990). Toward a feminist theory of caring. In E. Abel & M. Nelson (Eds.), *Circles of care* (pp. 36–54). SUNY Press.

van Valkengoed, A. M., & Steg, L. (2019). Meta-analyses of factors motivating climate change adaptation behaviour. *Nature Climate Change, 9*(2), 158–163.

Westley, F. R., & Antadze, N. (2010). Making a difference: Strategies for scaling social innovation for greater impact. *The Innovation Journal: The Public Sector Innovation Journal, 15*(2). http://innovation.cc/scholarly-style/westley2antadze2make_difference_final.pdf

Wetherell, M., McConville, A., & McCreanor, T. (2020). Defrosting the freezer and other acts of quiet resistance: Affective practice theory, everyday activism and affective dilemmas. *Qualitative Research in Psychology, 17*(1), 13–35.

Whyte, K. P. (2017). Our ancestors dystopia now: Indigenous conservation and the Anthropocene. In U. K. Heise, J. Christensen, & M. Neimann (Eds.), *The Routledge companion to the environmental humanities.* Routledge.

Wiesel, I., Steele, W., & Houston, D. (2020). Cities of care: Introduction to a special issue. *Cities, 105*, 1–3.

Williams, M. (2017). Care-full justice in the city. *Antipode, 49*, 821–839.

Williams, M. J. (2020). The possibility of care-full cities. *Cities, 98*, 1–7.

Wilson, J., & Swyngedouw, E. (Eds.). (2015). *The post-political and its discontents: Spaces of depoliticisation; spectres of radical politics.* Edinburgh University Press.

INDEX[1]

[1] Note: Page numbers followed by 'n' refer to notes.